T0155592

Programming LEGO® EV3 My Blocks

Teaching Concepts and Preparing for FLL® Competition

Gene Harding

Apress®

Programming LEGO® EV3 My Blocks: Teaching Concepts and Preparing for FLL® Competition

Gene Harding
South Bend, Indiana, USA

ISBN-13 (pbk): 978-1-4842-3437-2 ISBN-13 (electronic): 978-1-4842-3438-9

https://doi.org/10.1007/978-1-4842-3438-9

Library of Congress Control Number: 2018935947

Managing Director, Apress Media LLC: Welmoed Spahr
Acquisitions Editor: Aaron Black
Development Editor: James Markham
Coordinating Editor: Jessica Vakili

Cover designed by eStudioCalamar

Cover image designed by Freepik (www.freepik.com)

Distributed to the book trade worldwide by Springer Science + Business Media New York, 233 Spring Street, 6th Floor, New York, NY 10013. Phone 1-800-SPRINGER, fax (201) 348-4505, e-mail orders-ny@springer-sbm.com, or visit www.springeronline.com. Apress Media, LLC is a California LLC and the sole member (owner) is Springer Science + Business Media Finance Inc (SSBM Finance Inc). SSBM Finance Inc is a **Delaware** corporation.

For information on translations, please e-mail rights@apress.com, or visit http://www.apress.com/rights-permissions.

Apress titles may be purchased in bulk for academic, corporate, or promotional use. eBook versions and licenses are also available for most titles. For more information, reference our Print and eBook Bulk Sales web page at http://www.apress.com/bulk-sales.

Any source code or other supplementary material referenced by the author in this book is available to readers on GitHub via the book's product page, located at www.apress.com/978-1-4842-3437-2. For more detailed information, please visit http://www.apress.com/source-code.

Printed on acid-free paper

This book is dedicated generally to the coaches, mentors, and other volunteers who make FIRST® LEGO® League possible, and specifically to the 2015–2016 Titan Robotics team of students and their parent coaches.

Table of Contents

About the Author

Gene Harding is an Associate Professor of Electrical and Computer Engineering Technology at Purdue University, where he has taught since 2003. He has three years of industrial experience with Agilent Technologies, and 28 years of combined active and reserve service in the United States Air Force. He holds an MSEE from Rose-Hulman Institute of Technology, and is a licensed professional engineer. Professor Harding coached FIRST® LEGO® League teams in a highly competitive region for five years while his son was participating. In his final year of coaching, his team placed second in the state out of more than 300 teams, and qualified for the international tournament at Legoland, where his team won the Core Values Teamwork award and placed third in the Robot Game out of 72 teams from all over the world.

About the Technical Reviewer

Massimo Nardone has more than 22 years of experience in security, Web and mobile development, and cloud and IT architecture. His true IT passions are security and Android.

He has been programming and teaching how to program with Android, Perl, PHP, Java, VB, Python, C/C++, and MySQL for more than 20 years. He holds a Master of Science degree in Computing Science from the University of Salerno, Italy.

He has worked as a project manager, software engineer, research engineer, chief security architect, information security manager, PCI/SCADA auditor, and senior lead IT security/cloud/SCADA architect for many years. His technical skills include security, Android, cloud, Java, MySQL, Drupal, Cobol, Perl, Web and mobile development, MongoDB, D3, Joomla, Couchbase, C/C++, WebGL, Python, Pro Rails, Django CMS, Jekyll, Scratch, and more.

He currently works as Chief Information Security Officer for Cargotec Oyj. He previously worked as visiting lecturer and supervisor for exercises at the Networking Laboratory of the Helsinki University of Technology (Aalto University). He also holds four international patents (PKI, SIP, SAML, and Proxy areas).

Massimo has reviewed more than 40 IT books for different publishing companies and he is the coauthor of *Pro Android Games* (Apress, 2015).

Preface

Some background information might be useful for many before starting: First, I provide an overview of *FIRST*® LEGO® League (FLL®) for those who might not be familiar with it, followed by a description of the Robot Game, the purpose and focus of the book, and finally how the book is organized.

FLL® is an acronym for *FIRST*® LEGO® League. *FIRST*® is also an acronym: For Inspiration and Recognition of Science and Technology. *FIRST*® is comprised of four levels of robotics competitions: FLL® Jr., FLL®, *FIRST*® Tech Challenge (FTC), and *FIRST*® Robotics Challenge (FRC). Although the rules specify age limits instead of grade levels, in the United States (it varies in some other countries) FLL® Jr. is normally for third grade and below, FLL® for fourth to eighth grade, FTC for seventh to twelfth grade, and FRC for ninth through twelfth grade.

It is worth discussing how FLL® works because it is not intuitively obvious at first glance. There are four parts to competitions: three judging rooms and the Robot Game. It turns out that the Robot Game, which is the only part that spectators can see, is only a qualifier. It is also the most quantifiable of the four activities. The big award at FLL® tournaments is the Champion Award, which is won or lost in the judging rooms. The catch is that teams can only compete for the Champion Award if they qualify by finishing in the top 40 percent of the Robot Game. If they do not meet this criterion, they cannot compete for the Champion Award no matter how strong their judging room performances. The three judging rooms are Core Values, Project, and Robot Design.

One could think of the eight Core Values as character-building values. This is how they were worded when I was coaching:

- We are a team.
- We do the work to find solutions with guidance from our coaches and mentors.
- We know our coaches and mentors don't have all the answers; we learn together.
- We honor the spirit of friendly competition.

- What we discover is more important than what we win.
- We share our experiences with others.
- We display Gracious Professionalism® and Coopertition® in everything we do.
- We have FUN!

In competition, most teams do a good job at practicing the Core Values. The kids and coaches are great sports. Teams encourage and cheer for each other, even as they compete intensely to win (hence the terms Gracious Professionalism and Coopertition).

In the Core Values judging rooms, each team is presented a challenge, such as a puzzle or team-building exercise. The kids are observed during the challenge, then there is a question-and-answer (Q&A) period during which the judges ask questions. Team members are evaluated against a rubric that focuses more on how they interact than on how well they solve the challenge.

The Project is a major part of FLL. Each year there is a new theme. Recent themes have included Trash Trek℠, World Class℠ Learning Unleashed, Nature's Fury℠, and Senior Solutions℠. Within the parameters specified, each team must identify an existing problem related to the theme, research existing solutions, develop a new and innovative solution to solve the problem, and then compose a presentation to concisely communicate their solution to the judges. In the Project judging room, the five-minute presentation usually takes the form of a skit, after which there is a five-minute Q&A with the judges.

The Robot Design judging is linked directly to the Robot Game, specifically how the team approached the Game. They must explain to the judges their strategy, why they chose the missions they chose, how they did the programming and robot building, and so on. They are judged against a rubric that includes mechanical design, programming, and strategy and innovation, each with three subcategories.

Overall, FLL® is a tremendous program. Just think about all of the learning and development the kids experience. They learn how to design as they program and build their robot. In so doing they apply math and learn about physics concepts, such as center of gravity, torque, moment arms, gear reduction, friction, and traction. They do research and they write. They perform all manner of problem solving. They work together in teams. They practice drama as they prepare a skit and public speaking when they present to the judges. They learn to overcome adversity (there is no shortage of adversity in FLL competition, or the weeks leading up to it). In short, they prepare for life. I am a big believer in FLL.

The Robot Game is the part of competition that is open to the public, and is the most objective and quantifiable of the four components. The game is played out on a 4' × 8' table with an enclosing border that is approximately 3" high. On the table is the "playing field" mat, and affixed to the mat are various LEGO models representing the different missions teams can perform. The mat, models, and missions change each year to match that year's theme, so a kit must be purchased to create a standardized practice table to prepare for competition.

Each team can purchase no more than one Field Kit, which includes the mat and hundreds of LEGO pieces the kids use to assemble the field models. Detailed pictorial instructions to build the models are posted online. Building the models takes a substantial amount of time, but is an exercise most kids thoroughly enjoy. It is important that the models be built precisely according to the instructions, so it is wise to have the kids double-check each other, and one or more coaches or mentors should do a follow-up check after that to ensure each model was assembled correctly.

The kids design and build their own robot, which must be built entirely of genuine LEGO parts. The core of the robot is called the *brick*.

The third-generation brick is the EV3. Its heart is an ARM9™-based microprocessor chip, the Sitara AM1808, made by Texas Instruments (TI). The AM1808 has 128 kB of on-chip memory, quite a variety of peripheral interfaces, and Linux support. The EV3 also has Bluetooth capability provided by TI's CC2560 Bluetooth controller. It has six control and input buttons, an LCD display that is approximately 1.1" × 1.7", four ports to drive motors and monitor motor rotation, four ports to monitor sensor data, and a Mini-B USB receptacle for interfacing with a PC that has the EV3 software installed. It also has a Type A USB receptacle and SD card port. The motor and sensor ports use cables and connectors that are very similar to the old RJ-11 telephone cables. A LEGO kit with a brick normally comes with a USB cable with a Mini-B plug on one end for connection to the brick and a Type A plug on the other end for connection to a PC with a USB port. It is used primarily to download EV3 programs to the brick, including firmware updates, and to manage its memory.

The brick comes with provision for six AA batteries, but a rechargeable battery can be purchased for it. This option is highly recommended. Without it, the brick must be removed from the robot to replace the batteries, and this can be a very tedious process with many robots. If the rechargeable battery is installed, there is also a port to recharge it.

As long as a small space is left to plug in the cord, recharging the battery is very easy, and disassembling the robot is not necessary.

There are very few constraints on the robot design, resulting in a wide variety of designs at FLL competitions. The robot must, however, start completely within base, which is normally an area between one and two square feet at the southwest corner of the table. Once the robot leaves the base, it must be autonomous. Team members may not touch the robot while it is out of base without receiving a penalty. Competition Robot Game rounds last 2.5 minutes. Most teams try to do three to five mission sets with their robot during the allotted time. Two team members may be at the table at any given time, and they can swap out during the 2.5 minutes, much like soccer players or professional wrestlers. There are three 2.5-minute rounds to the Robot Game, with time between rounds to make adjustments. The highest score of the three rounds is the one counted.

A typical run goes something like this: The team gets their robot ready to start the round, gets the first program queued up, positions the robot, and waits for the "go" signal. The announcer says "3, 2, 1, LEGO!" and one of the team members presses the button to start the program. The team members wait and watch as their robot navigates around the table performing one or more missions. If it fails to return to base properly they retrieve it and take a penalty. Either way, once it is back in base they change attachments to reconfigure the robot, queue the next program, position the robot, and press the button to execute the next program. The process repeats until either all of the (planned) missions are done, or the 2.5-minute timer expires.

To perform these missions, the kids must build special attachments for the robot, also comprised entirely of LEGO parts, and write programs to control the robot using the EV3 software, a graphical programming language created by National Instruments.

The motivation for writing this book was twofold: first, to assist coaches and mentors—especially those with little or no programming experience—in preparing their teams for FLL competition; second, to assist anyone interested in teaching robotics and programming concepts using the EV3. It is possible that teachers at the primary, secondary, and even postsecondary levels might want to use the EV3 for instructional purposes. For that reason, I have tried to carefully explain the theory and algorithms behind the various programs.

It turns out that there are a number of common tasks that must be accomplished every season, regardless of the specific layout of the field mat. Such tasks include moving forward and backward, turning, and finding and following lines. Subroutines to perform these tasks can be written and debugged during the off-season time, facilitating more rapid and precise program development during the season after the Robot Game specifics are released. These subroutines are called My Blocks in LEGO-speak, and they are the focus of this book.

The book is organized as follows. Chapter 1 discusses some of these common tasks as a preview of what is coming in Chapters 3 through 7. It also discusses a number of concepts that are key to successful programming and execution on the game board. For completeness, Chapter 2 gives a brief introduction to EV3 programming, although it should be noted that there are numerous Internet sites that do the same. Chapter 3 discusses moving the robot forward and backward, Chapter 4 covers turning in place, Chapter 5 examines using the wide black lines on the mat for precise navigation, Chapter 6 looks at using table walls for navigation, and Chapter 7 touches on advanced topics. Chapter 8 closes with some final thoughts about preparing for competition. There is also a glossary in the back with definitions for a number of key terms.

Chapters 3 through 7 each begin with a list of learning topics covered in the chapter. Each topic that is new to the chapter is covered when appropriate in the chapter. If the topic was introduced in an earlier chapter, a cross-reference is given when it is revisited.

Some of the topics, such as moving forward, can be done with different levels of complexity. Although more complex approaches can offer big advantages, they are harder to understand and take longer to program, debug, and tune. These more advanced topics are presented in Chapter 7 for those who are interested.

The programming sections include numerous screenshots and step-by-step instructions to walk through the process of implementing each program.

There are many opportunities to volunteer for FIRST as a coach, mentor, referee, judge, and so on. If you are interested, see `https://www.firstinspires.org/` for more info.

I am creating a companion web page to supplement this book. Look for Mr. G's EV3 page at `https://polytechnic.purdue.edu/profile/glhardin/MrGEV3`.

I wish you the best in your programming endeavors, whether your goal is victory in FLL competition, teaching your students, learning for yourself, or all of the above!

Acknowledgments

First thanks must go to Ed B. and Laura G., who got me started in FLL by asking me to help them coach The Bricks. Ed was the one who first encouraged me to take on the role of head coach. Daryl R. was a great help as the Core Values coach on that team.

When my son moved to a new school, Jeff K. and I helped Wendy C. create a new team at Trinity School at Greenlawn. The Head of School, John L., wanted us to start a team. He and the IT director, Mark H., plus one or more anonymous donors, were very supportive in providing us with a good space and equipment to start things off. Philip D. was also a huge help, building a wonderfully made table for practicing the Robot Game.

In my second year at Trinity, I again took on the role of head coach, but had tremendous help from several other parents. Jeff K. again helped, with both the Robot Game and Project. Pat M. and Blake L. stepped up in a huge way to support the kids in development of the Project. Amy D. also provided a lot of support there. Frank R. was an amazing Core Values coach. John and Claire K. stepped in at various points to help the effort. Coaching Team 1115, Titan Robotics, was truly a team effort, the best I have seen firsthand.

Next, I want to thank the kids that formed the team: Helena D., Graham H., Jackson K., Ceci K., Peter M., Lupé P., and Peter R. Unfortunately, Lupé had to leave before the qualifying tournament. We were sad to see her go. Peter M. was a key member of the team through the state tournament, where we finished second. The other five hung in there for five more months, totally overhauling the Robot Game as well as making numerous updates to the judging materials to prepare for the North American Invitational Tournament at Legoland in Carlsbad, CA. Their efforts paid off with a third-place finish in the Robot Game and first place in the Core Values Teamwork category. What a ride! It was an amazing team and a wonderful, although at times grueling, experience. Thanks to all of them!

Finally, I want to thank those who reviewed some or all of the book to make it better. Cyndi A. and her husband were my first reviewers, providing great feedback on the first two chapters. Brad S. was amazing, faithfully reviewing the entire book and providing great insight along the way. Scott C. did not actually review the text, but gave me some crucial feedback that allowed me to dramatically improve and simplify Chapter 3. Last but not least, my son Graham reviewed the toughest portions of Chapter 7.

CHAPTER 1

Concepts and Common Tasks

There are several concepts that should be covered to lay the foundation for the heart of this book, including subroutines, feedback, troubleshooting, motor and color sensor matching, and unit conversion, among others.

My Blocks, or Subroutines

In programming, a subroutine is a program that is used to do a task that must be done over and over. Rather than doing the programming for such a task many times, a "mini" program called a subroutine is written just to do that task. The program is carefully debugged and tuned, then it is "called" and run by programs that need it, avoiding the need to repetitively create and debug the code. The principle of dividing programs into smaller, more manageable blocks is called *modularity,* and it can be a tremendous time saver. The term used for such subroutines in LEGO®/MINDSTORMS®/EV3/NXT programming is "My Block."

So, are there times and places in *FIRST*® LEGO® League (FLL®) where it makes sense to use My Blocks? Absolutely! In fact, there are tasks that must be done every season by every robot, regardless of the Robot Game mat layout, missions, or LEGO models used on the mat. Common tasks include moving forward and backward a specified distance, in either a straight or curved line; turning in place; and finding, following, and squaring up on lines. Because these tasks are useful for any mat layout, they can be worked on during the off-season, before the Challenge guide is published for the coming season. Having a functional library of My Blocks ready to go at the start of the season allows better focus on the unique aspects of that year's Robot Game. My Blocks are covered in more detail in a later chapter.

G. Harding, *Programming LEGO® EV3 My Blocks*, https://doi.org/10.1007/978-1-4842-3438-9_1

Feedback

Another important concept in more advanced programming is feedback. Feedback is the information that is monitored, or "fed back," to the control algorithm. The feedback is compared to the desired value, and differences indicate how to make corrections to get closer to the desired value. For example, when following a line, the program must constantly monitor light intensity to track the line accurately. Feedback information typically comes from one of the sensors, such as the color sensor, motor rotation sensor(s), gyro sensor, ultrasonic sensor, and touch sensor. The color sensor is invaluable for finding and following lines, and the motor rotation sensors are important for determining distance traveled. The ultrasonic sensor can be used to help a robot follow smooth, regular surfaces like the game table walls, and to determine distances to objects. The gyro sensor can be useful for determining which direction the robot is facing, and the touch sensor can be used to determine when the robot has contacted an object. The EV3 brick has four sensor ports. Motor rotation sensor information is transmitted through the motor ports, leaving all four sensor ports available for other sensors. Specific sensor usage and associated feedback techniques are covered in more detail in later chapters.

The preceding paragraph discussed using feedback to allow the robot to monitor itself and make corrections for more precise, reliable operation, but there is another type of feedback: user feedback. This type of feedback is not for the robot, but for the person(s) running the robot. For example, when you are troubleshooting it can be invaluable to know that the robot actually detected the line it just passed, or that it just activated its steering correction to maintain a straight course. Such feedback can be provided by changing the color of the lights on top of the brick, or by providing an audio signal, such as a beep, to cue the user without requiring the robot to stop mid-mission. User feedback is covered in more detail in a subsequent chapter.

Troubleshooting

Troubleshooting is an area that can be very tricky. It is often very tempting, after observing some errant behavior in the robot, to make a premature conclusion rather than thinking through the situation carefully. If the program is then changed to "fix" the problem, it can end up introducing more problems instead of improving the program. There are a couple of things one can do avoid this. First, slow down and think carefully about the underlying problem that is causing the errant behavior; second, check to make sure you can consistently re-create the problem. Doing so takes discipline, but can save a lot of grief caused by erroneous "corrections."

Unit Conversion

A concept that is very helpful for making My Blocks more usable is unit conversion. For instance, suppose you want the robot to go forward 3", and your robot has wheels with a 10" circumference. You could simply use a Move Steering block, choose the mode "On for Rotations," and set the Rotations parameter to 0.3 (Figure 1-1). You could also choose the mode "On for Degrees" and set the Degrees parameter to 108 (Figure 1-2). The problem with this approach is that it requires the programmer to do a calculation every time the Move block is used. Because the EV3 software contains Math blocks, why not use them to do the calculations for you? A Math block can easily convert inches to either rotations or degrees, so embedding the Math block conversion with a Move block inside a My Block can result in quicker and more accurate programming. As the calculations become more complex, the reliability and consistency of a debugged My Block become very useful. This concept is explored in more detail in later chapters.

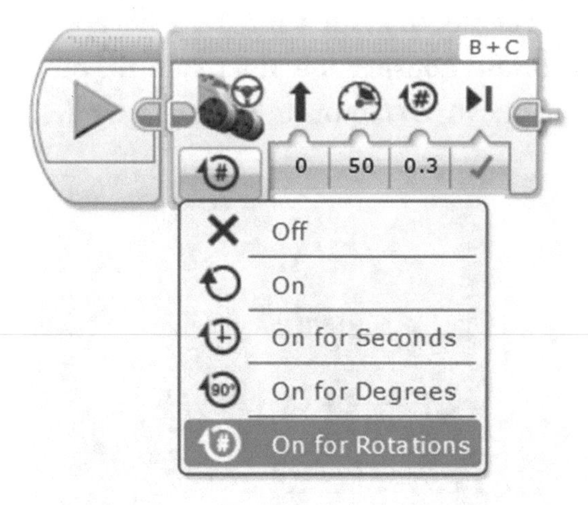

Figure 1-1. *Move Steering, 0.3 wheel rotation*

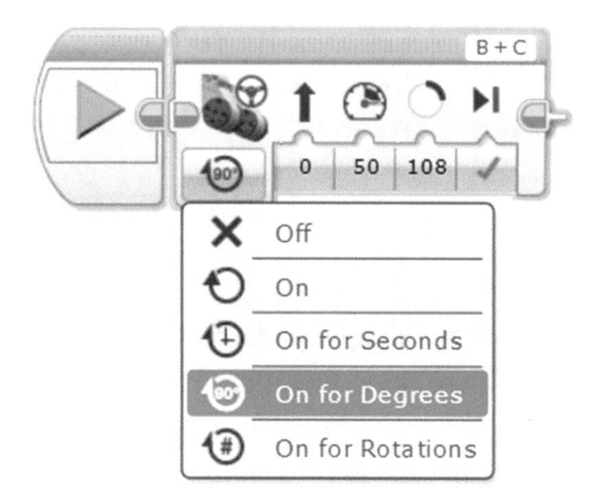

Figure 1-2. *Move Steering, 108° wheel rotation*

Motor and Color Sensor Matching

A typical robot uses two drive motors. If the motors do not turn the same number of rotations or degrees with the same input power, the robot will not travel in a straight line. We saw this problem more with the NXT than with the EV3, but there are ways to find motors that are well matched to minimize this problem. The technique we employed uses a differential, as shown in Figure 1-3. Set up the motors so that they drive the differential in opposite directions, then run a program that applies the same power level to each motor. If the differential housing does not rotate, the motors are matched. If it rotates in either direction, one of the motors is turning faster than the other.

Figure 1-3. *Example differentials*

Similarly, if your robot uses two color sensors, it helps to use two sensors that are closely matched. This can be determined by observing the baseline values of the sensor during calibration. When calibrating one of the sensors, note the minimum and maximum values for the calibration, then repeat for the other sensor. If they are close, the sensors are well matched. If not, it would be wise to find a pair of sensors that are matched. The reason for this is that the calibration settings for the sensor that is being calibrated are applied to all of the color sensors connected to the brick, so if the sensors are not matched the results from one sensor might not match results from the other sensor(s). (Color sensor calibration is covered in Chapter 5.)

Levels of Complexity

Many tasks can be performed at different levels of complexity and effectiveness, allowing beginners to create effective programs, then improve them as they gain experience. For instance, a basic My Block for moving forward a specified distance might just convert inches to degrees of rotation. A more advanced version could monitor a gyro sensor and add steering correction to ensure the robot moves in a straight line. An even more advanced My Block might incorporate acceleration and deceleration to allow for higher speed without wheel slip. (Wheel slip introduces error, making the distance moved inaccurate and inconsistent.)

Commenting and Program Development

Commenting programs is very important. It is a big help to others who need to use or modify the program. Comments can also be very helpful, even for the original programmer, if he or she has not looked at the program for some time. In FLL competition, programs are often written and worked on as teams, so comments can be invaluable for effective programming. They are also useful for judges evaluating the programs, and demonstrate a higher level of professionalism than programs without comments.

In addition to commenting, a good practice is to build up complex programs a few blocks at a time, then debug and tune before continuing. Debugging and tuning are generally quicker and easier when done in smaller chunks. An additional benefit is that the earlier parts of the program are run many times and typically end up being thoroughly tested. Running early parts of the program many times also has a downside. For large programs, a lot of time can be "wasted" rerunning parts of the program that have already been tested and verified. This issue can often be mitigated by dividing the program into parts during development, then later combining the parts to make the larger program.

Errors and Consistency

There are two basic types of errors: random and systematic. Random errors are caused by limitations of the robot itself or the environment in which it operates. They are generally beyond the control of the user, although there are sometimes ways to minimize their effects. Systematic errors are caused by inaccurate calibration or other mistakes in the robot's construction or programming. Consider the example of a My Block programmed to input a distance in inches, convert it to degrees of wheel rotation, and move the robot forward the specified distance. Suppose the robot was programmed to move forward 10" and, after running the program several times, the robot stopped at distances ranging from 10.75" to 11.25", with an average distance of 11.00". The measurements imply a systematic error of 11.00" - 10.00" = 1.00". This can be corrected by adjusting the Math block to reduce the conversion factor by 9 percent so that the robot travels 10" instead of 11".

On the other hand, it appears that the robot's travel has a random error of ±0.25", so it can consistently hit 10.00" within 0.25", but could end up anywhere between 9.75" and 10.25" (once the conversion factor is calibrated properly). This might be a limitation of the robot itself. Moreover, this error is likely to scale with distance, so it may be more than 0.25" at longer distances.

For Titan Robotics, the last team I coached, we established a "five of five," or "5/5" rule: Our goal was to run each mission successfully five times out of five tries. To achieve that level of consistency (not that we always succeeded), it is important to eliminate systematic errors and compensate for random errors.

Resetting the EV3 Brick

Occasionally, the EV3 brick might "lock up" and fail to respond to any inputs. Although removing the battery pack will provide a hard reset to restart the brick, this option can be very inconvenient because it often requires substantial disassembly of the robot. Fortunately, there is an easy way to reset the brick (see Figure 1-4):

1. Press and hold the Back, Left, and Center buttons.
2. When the display goes blank, release the Back button.
3. When the display says "Starting," release the Left and Center buttons.
4. Wait for the brick to finish its normal boot process.

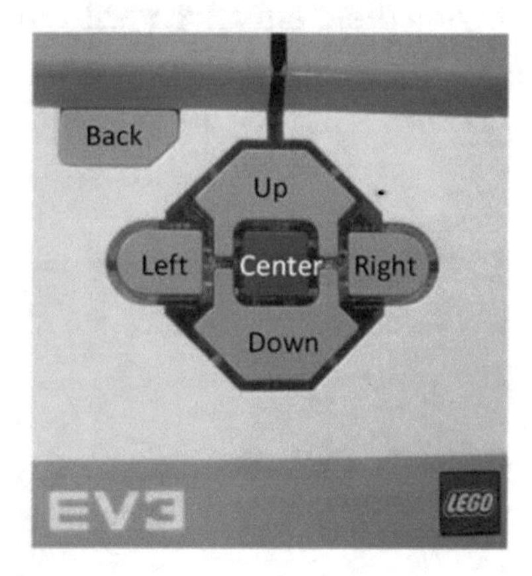

Figure 1-4. *EV3 brick buttons*

Updating the Firmware

The term "firmware" refers to software that is stored in read-only memory. It is a quasi-permanent type of software, generally designed to facilitate interaction with the hardware in a computer system. Although infrequent, LEGO does occasionally publish firmware updates to fix bugs or provide enhanced functionality in the brick. The Home Edition of the firmware can be downloaded at `www.LEGO.com`, and the Education Edition from `education.LEGO.com`.

When updates are published, it is probably a good idea to update your brick's firmware, although if your robot is functioning reliably and a competition is approaching, it might be better to wait. If you do update and have problems, you can revert to the previous version if you kept a copy. Also, if your robot begins acting erratically, and nothing seems to fix it, a firmware update to "reset" the firmware is worth a try. To update the firmware, make sure the brick is on and connected to your PC, then go to the Tools menu and select Firmware Update.

Conclusion

This chapter introduced a number of concepts you can teach your kids as they learn how to program the EV3 brick. These concepts are used by engineers and scientists on a daily basis. Learning and applying them effectively will contribute to success in FLL competition, but it will also leave them with skills they can apply in many areas as they grow older.

CHAPTER 2

Getting Started with the EV3

There are numerous web pages online that provide an introduction to EV3 hardware and software, but for the sake of completeness this chapter provides such an introduction to help beginners get started, and share a few tips that might be useful even for some experienced users. There are two different types of EV3 kits that can be purchased, so I will begin with a description of the similarities and differences between the two kits, including variations in both hardware and software. Next is a description of the user interface to explain file management, communicating with the brick, creating programs, and operating the brick. The chapter finishes with step-by-step instructions to create and run a couple of basic programs.

Education vs. Home Editions

There are two versions of the EV3 kit: Education and Home Editions. In terms of hardware, both kits include one brick, two large motors, one medium motor, and one color sensor. The Education Edition also has a second touch sensor, a gyro sensor, and an ultrasonic sensor. The Home Edition has an infrared sensor and associated remote control, which allow the robot to be controlled remotely. The hardware differences are summarized in Table 2-1.

G. Harding, *Programming LEGO® EV3 My Blocks*, https://doi.org/10.1007/978-1-4842-3438-9_2

Table 2-1. *Key Hardware Features for Education and Home Editions*

Item/Feature	Education Edition	Home Edition
Brick	1	1
Large motor	2	2
Medium motor	1	1
Color sensor	1	1
Touch sensor	2	1
Gyro sensor	1	0
Ultrasonic sensor	1	0
Infrared sensor	0	1
Infrared remote control	0	1
Rechargeable battery and charger	1	0

There are also two editions of the EV3 software. The Education Edition has the ability to do experiments as well as programs, and can do data logging (experiments and data logging are beyond the scope of this book). The Home Edition also does not come with software support for gyro and ultrasonic sensors, but software blocks can be added to support those sensors. The sensors can be purchased directly from LEGO or from many other places online. The software blocks are gyro.ev3b and ultrasonic.ev3b, respectively, and are available for free download at `www.LEGO.com`.

The Programming Interface

The EV3 software opens in a window called the Lobby. Although the Education Edition Lobby and Home Edition Lobby look different, both have a conventional top menu bar that can be used to create new projects and open existing projects, as shown in Figure 2-1.

Figure 2-1. *Education Edition Lobby (top) vs. Home Edition Lobby (bottom)*

To open an existing project in either edition, from the menu bar select **File ➤ Open Project...**. Let's create a new project: From the menu bar select **File ➤ New Project**. (In the Education Edition you must additionally select **Program** instead of **Experiment**.) This opens the programming window shown in Figure 2-2. The Content Editor is outside the scope of this book, so close it by clicking on the icon, on the right side of the page near the top.

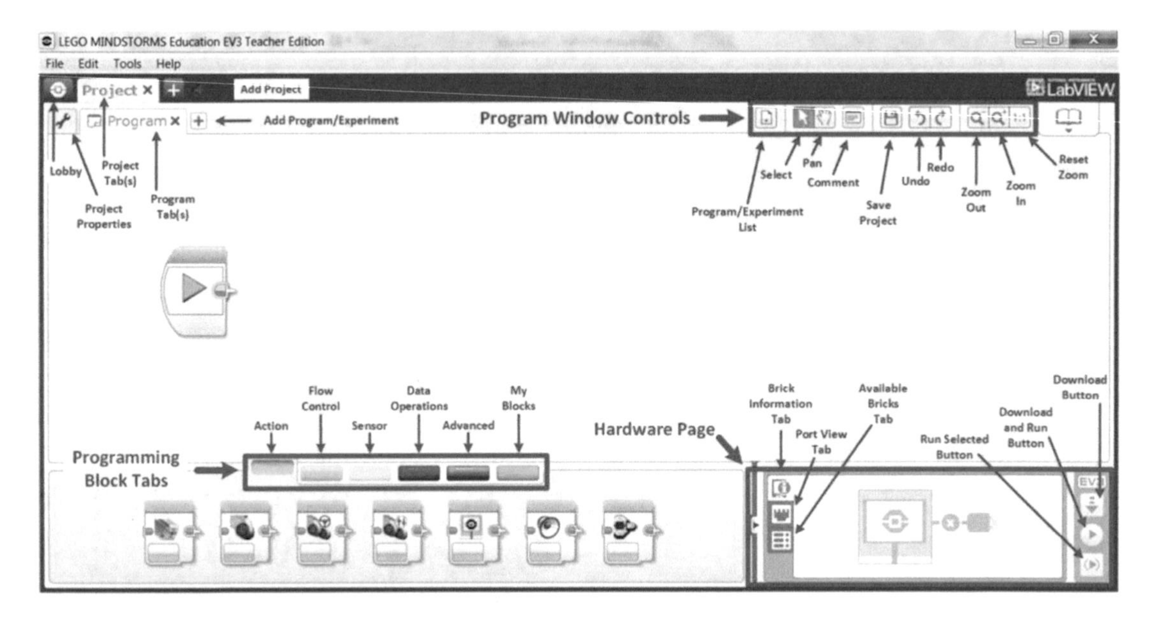

Figure 2-2. *EV3 programming window*

The row of buttons at the top right contains program window controls. (The program window is the space used to compose a program.) The first button provides a drop-down list of all programs in the current project, including My Blocks. The second button provides an arrow cursor for selecting blocks in the program window. A block can be selected individually by clicking on it, or multiple blocks can be selected by drawing a box around them. The Pan button enables moving the entire program left, right, up, or down in the display window. The Comment button places a comment window in the upper left corner of the program window. Once there it can be filled in, resized, and moved around as needed. The Save, Undo, Redo, Zoom Out, Zoom In, and Reset Zoom buttons function as their names imply.

The lower right portion of the display is called the Hardware Page. The page is unavailable, as shown in Figure 2-2, until a brick is connected to the computer. At the left of the Hardware Page are three tabs that control what information is viewed in the page: Brick Information, Port View, and Available Bricks. Three buttons are on the right side. The top button, Download, downloads the entire project to the brick. The middle button, Download and Run, downloads the project and runs the program currently shown in the Program window. The bottom button, Run Selected, downloads and runs only the blocks currently selected in the Program window. This option can be very handy for tuning programs.

Programming blocks are arranged in six tabs at the bottom left and middle: green for Action blocks, orange for Flow Control, yellow for Sensor, red for Data Operations, dark blue for Advanced, and turquoise for My Blocks.

At the top left part of the window is a traditional menu bar with File, Edit, Tools, and Help options. Below that are tabs to access the Lobby and each of the open projects, as well as to add a new project. Below those are tabs to access Project Properties and each program or experiment within the current project, as well as add a new program or experiment to the current project.

Clicking the Project Properties tab opens a new window, as shown in Figure 2-3. The lower portion of the page includes tabs for working with programs, images, sounds, My Blocks, variables, and exportable items. Buttons at the bottom allow the user to copy, paste, delete, import, and export items. One note of caution when deleting items: There is not pop-up window to confirm a deletion before it is deleted, so be very careful when deleting items. Once that button is clicked, the selected item is gone!

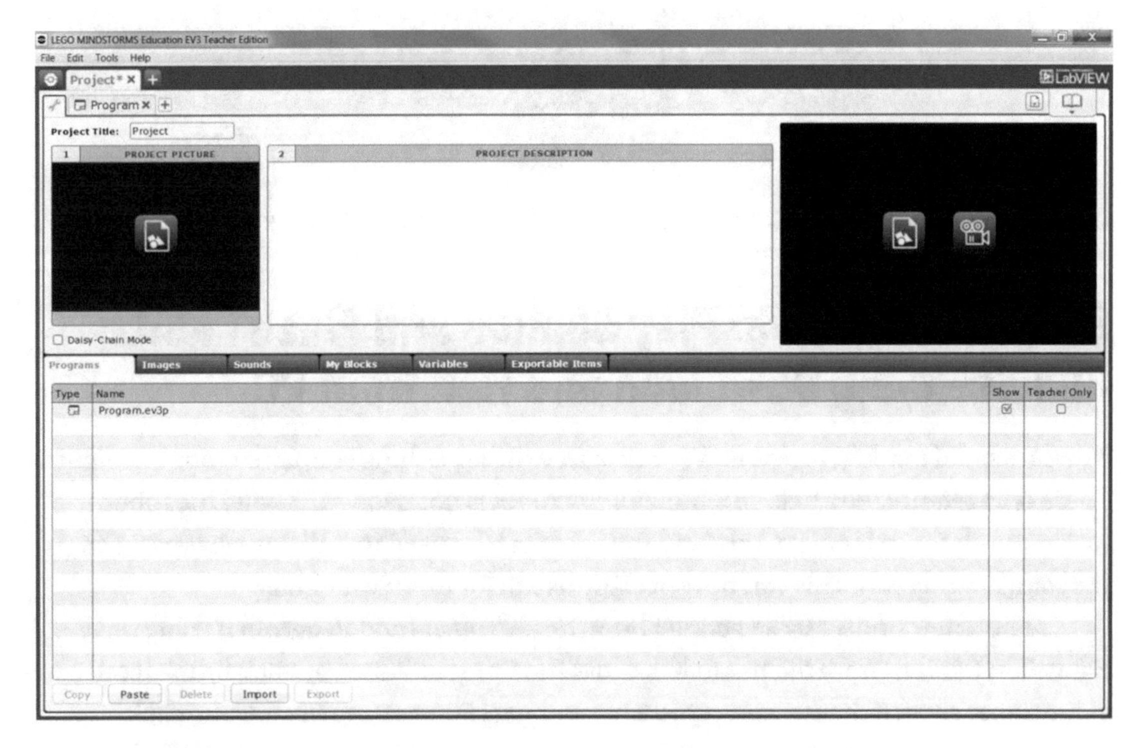

Figure 2-3. *Project Properties window*

One more item needs to be covered before we begin programming: data wires. Data wires are the virtual connections that allow information to be passed between blocks in a program. There are five types: text, numeric, numeric array, logic, and logic array. Array operations are beyond the scope of this book, but text, numeric, and logic data wires can be useful in various types of programming. The type of wire is indicated by its color and by the shape of the tab in the programming block: orange with a square tab for text, yellow with a rounded tab for numeric, and green with a pointed tab for logic. Figure 2-4 shows a notional program with each type of wire marked by an EV3 comment box. Notice that, in the second and third green Display blocks, the input tabs are square even though the data wires are numeric and logic, respectively. The data conversions from numeric to text and from logic to text are done automatically by the EV3 software. This is a change from the NXT, which was the second-generation Mindstorms software. With the NXT, a separate block was necessary to explicitly perform data conversions.

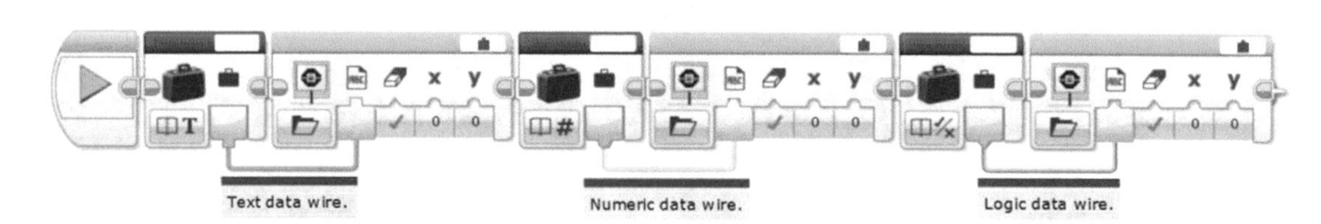

Figure 2-4. *Text, numeric, and logic data wire types indicated by color*

Simple Program to Play Sounds and Flash Lights; Run Selected, Download and Run from PC

Now let's build our first program. We will start with something simple: sounds and lights. These only require the brick, so a functioning robot is not needed. Having the robot make sounds is fun for most kids, and the lights can be used to indicate what part of a program is being run. This can be especially useful for troubleshooting.

Create a new project by selecting **File ➤ New Project** (and **Program** if using the Education Edition) from the menu bar. Close the Content Editor by clicking the icon at the upper right. Rename your project by saving it (**File ➤ Save Project As**, click the Save Project icon in the Program Window Controls area, or press Ctrl+Shift+S) as **MyFirstPrograms**. Double-click the **Program** tab and rename it **LightsAndSounds**. The programming window should look similar to Figure 2-5.

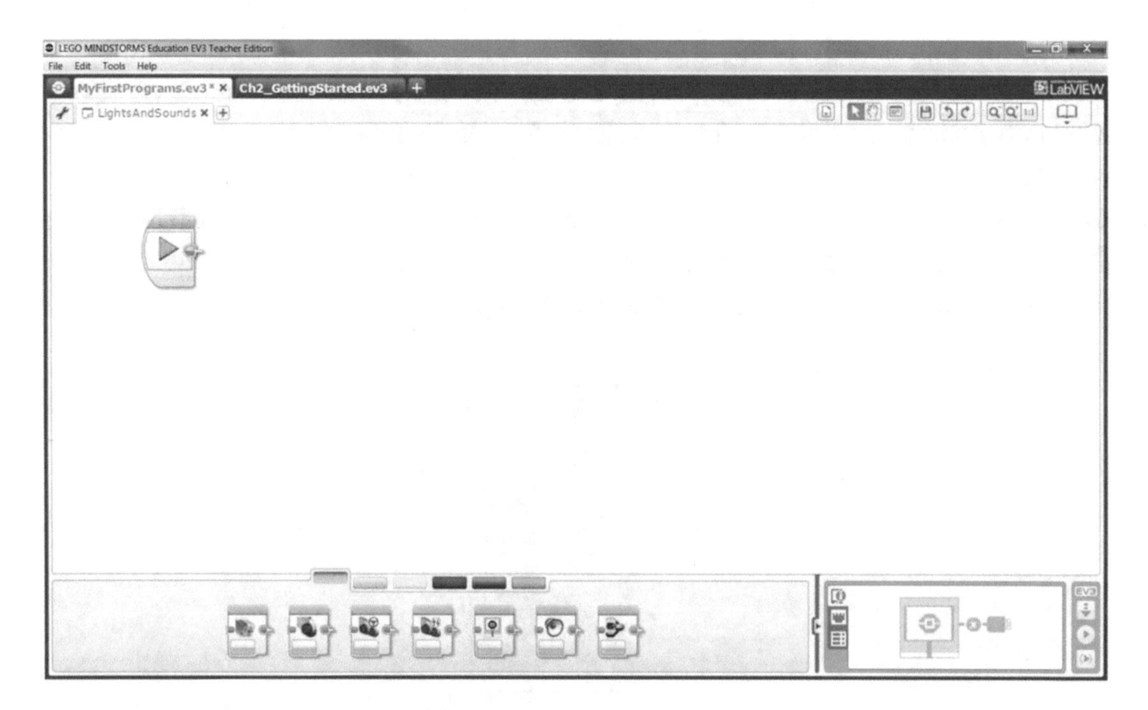

Figure 2-5. *Initial programming window for LightsAndSounds program*

The rightmost programming block under the green tab is Brick Status Light. Click on it and drag it onto the program window. Notice that, as it gets close to the Start block (the one already in the window), a gray "shadow" the shape of the Brick Status Light appears right next to the Start block, as shown in Figure 2-6. This "shadow" indicates where the block will "drop" onto the program chain when the left mouse button is released. Drop it there.

Figure 2-6. *Shadow indicating where block will drop onto program chain*

The first button in the block, at the lower left, is the mode button. Clicking it will give options (for this type of block) of **Off**, **On**, and **Reset**, as shown in Figure 2-7. Leave it set to **On**. The next button is a numeric input for selecting the light color. Clicking it reveals that **0** indicates a green light, **1** an orange light, and **2** a red light. Leave it set to **1** for orange. The third button is a logic input to determine whether the light will be pulsed (**True**) or not (**False** will leave the light on continuously). **True** is indicated by a check mark and **False** by an X. Change it to **False** so the light will be on continuously.

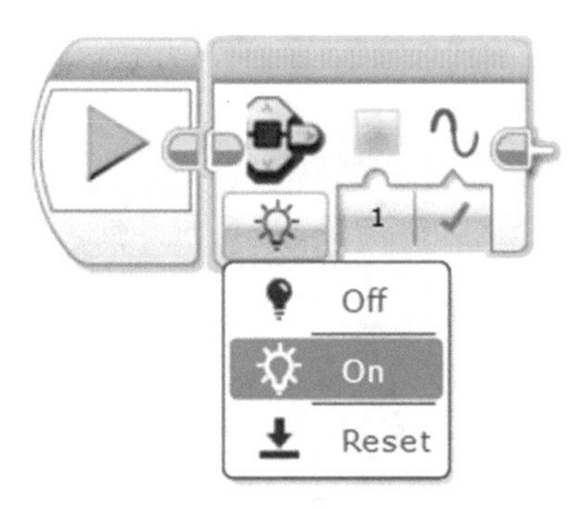

Figure 2-7. *Mode options for Brick Status Light block*

Next, place a **Sound** block onto the program chain right after the Brick Status Light block. Note that floating the mouse cursor over a block before you pull it into the programming window reveals its name, as shown in Figure 2-8. The **Sound** block's modes are **Stop**, **Play File**, **Play Tone**, and **Play Note**. It also has options for setting the **Volume** from 0-100 percent, and numeric options for **Play Type**. Leave **Volume** at 100 and **Play Type** at 0 (**Wait for Completion**).

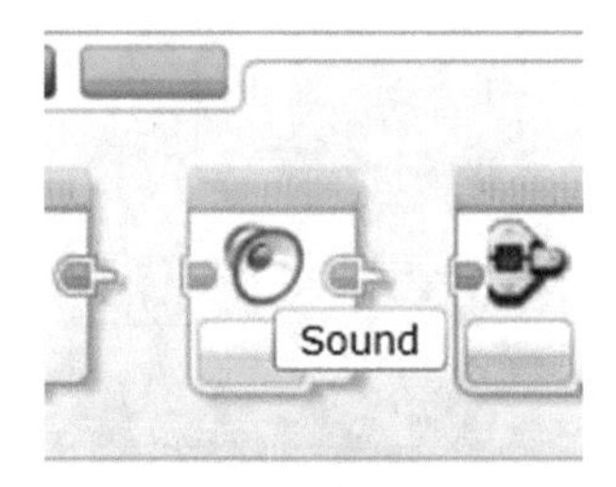

Figure 2-8. *Floating the cursor over a Sound block reveals its name*

Click in the white box in the upper right part of the Sound block to select a **File Name**. Although orange is not one of the predefined color options, yellow is, so select that by picking **LEGO Sound Files ➤ Colors ➤ Yellow** as shown in Figure 2-9. When you select **Yellow** you should also hear the sound played through the computer's speaker(s).

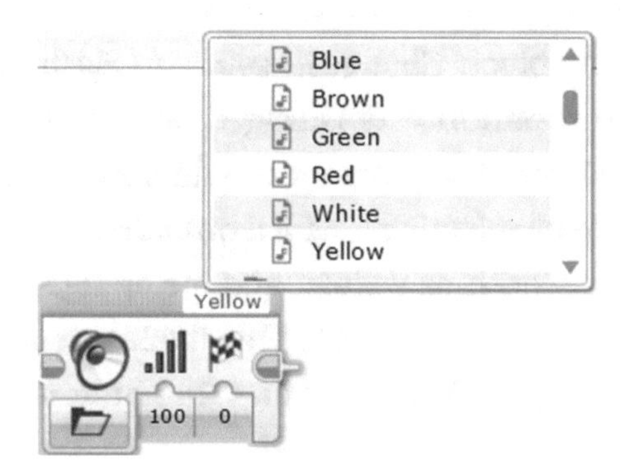

Figure 2-9. *Yellow Sound block*

Let's run this program using the **Download and Run** button. Turn on the brick by pressing the middle (dark gray) button. Connect the brick to the computer using the USB cable. The brick is ready when the status light turns green and it plays a brief tune called the "Download" sound (in the LEGO Sound Files, System folder). Click the **Download and Run** button (on the lower right, as shown in Figure 2-2). The program should download and run, as indicated by the status light briefly turning orange and the brick saying "yellow". The light goes out as soon as the Sound block is finished, which is fairly quick, so let's add a Wait block to make the light stay on longer. Go to the orange tab on the programming palette and drag a Wait block to the end of the program. Select mode **Time** and **2** seconds, as shown in Figure 2-10.

Figure 2-10. *Add 2-second Wait block*

Now let's add some blocks to see what the light looks like when it is pulsing. Draw a box that touches the right three blocks (Brick Status Light, Sound, and Wait) to select them. If you have selected them successfully, the block borders will turn turquoise. Press **Ctrl+C** to copy the blocks and **Ctrl+V** to paste them, then click and drag them to the end of the program chain. Copy the Yellow Sound block and insert the copy just after the second Brick Status Light block. Change its sound to Flashing by clicking the Yellow file name and selecting **Information ➤ Flashing**. Change the second Brick Status Light block's Pulse parameter to **True**. The program should now look like Figure 2-11. Verify the new part of the program by selecting it with a box and clicking the **Run Selected** button. The brick should say "flashing yellow" and the orange light should flash for a little more than 2 seconds (2 seconds plus the time it takes to play the audio). Now run the whole program by clicking the **Download and Run** button.

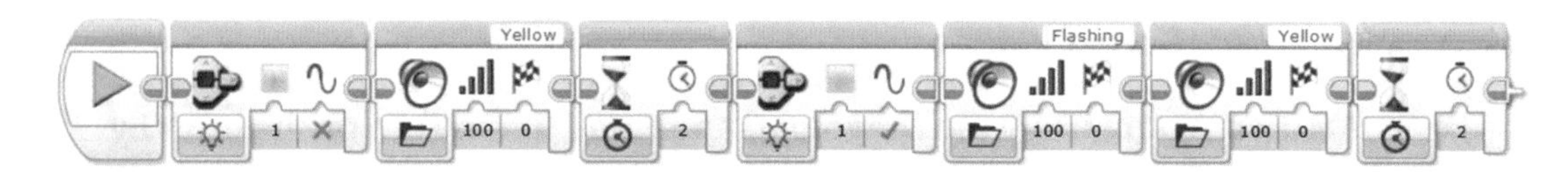

Figure 2-11. *Continuous yellow followed by flashing yellow*

Commenting is an important part of documenting programs, so let's add some comments. Click the **Comment** button in the Program Window Controls at the upper right. A comment window will appear at the upper left in the Programming window. Double-click in the window and type something like "Turn orange light on continuously, say "yellow" at full volume, and wait 2 seconds." Then resize the box to stretch across the second through fourth blocks and place the comment just above them. Repeat the process to comment the last four blocks, and the program should look like Figure 2-12.

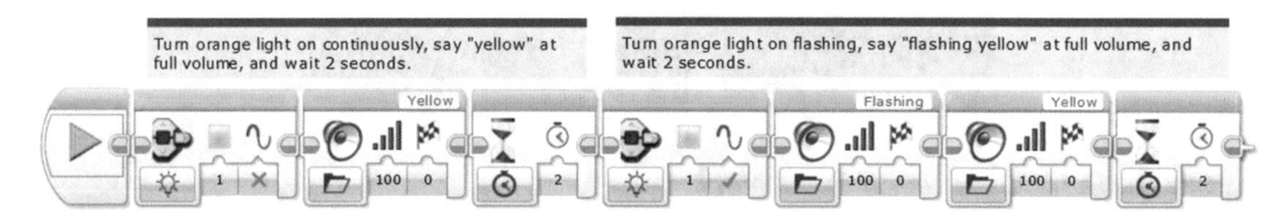

Figure 2-12. *First programming line with comments*

Note that there is another method for adding comments to programs: the Comment block, located under the Advanced programming blocks tab. It has one advantage over the comment boxes: Because it is a Program block it moves with the code when new blocks are added, whereas the comment boxes must be manually moved when programming changes reposition the programming blocks. On the other hand, the Comment block has a significant disadvantage: It is a fixed size, so it can only display a few dozen characters (three lines). Moreover, it cannot be placed over several blocks to provide a "summary" comment as shown in Figure 2-12.

Next, let's add sections to the program to repeat the process for the green and red lights. We will use copy and paste to make the process easier (press Ctrl+C to copy blocks, Ctrl+V to paste). Also, if we add sections directly to the end of the current program, the listing will extend off of the screen. This is okay, but harder to read, so we will add the new sections in the white space below the current program blocks. Select all of the current program, including comments, except for the Start block, then copy and paste them under the current program. Note that they are faded in color because they are not yet active parts of the program (see Figure 2-13). They must be connected to the first program line. Do this by clicking the Sequence Plug Exit on the right side of the last Wait block in the first line, then dragging a Sequence Wire and plugging it into the Sequence Plug Entry at the beginning of the second program line. This is illustrated in Figure 2-13. After you do this, the second program line will be displayed in bright colors like the first line.

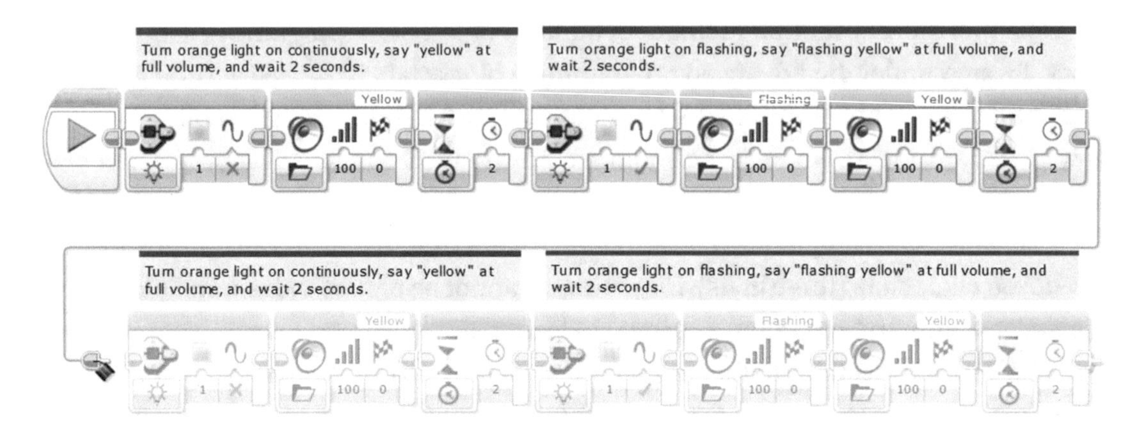

Figure 2-13. *Connecting program lines*

Now make the following changes to the programming blocks in the second line: (1) Change both Brick Status Light colors to **0** (green), (2) change both Yellow Sound blocks to **Green**, and (3) replace "orange" and "yellow" in the comment blocks with "green". Repeat the process for a third programming line, replacing "green" with "red" as shown in Figure 2-14. Run the program by clicking the **Download and Run** button. If interested, experiment with the different sounds, light options, and timing.

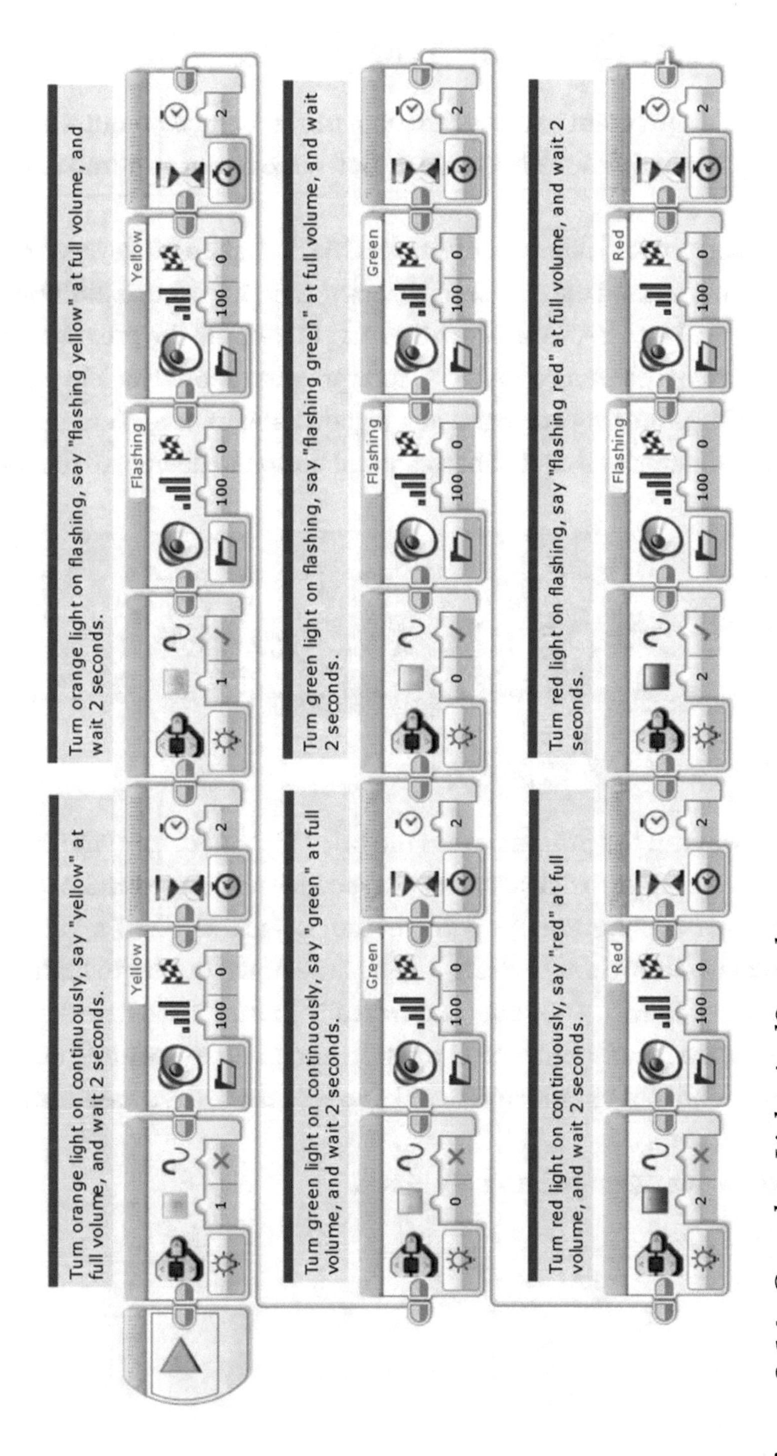

Figure 2-14. *Complete LightsAndSounds program*

Simple Program to Move Forward and Backward; Download from PC, Run from Brick

It's time now to write a program to make the robot move. This will require a working robot. The convention in this book is to wire the left large motor to port B and the right large motor to port C.

Open a new program and name it ForwardAndBack. Place a Move Steering block onto the program line and set the **Power** to 30. Leave Steering set to 0, **Rotations** to 1, and **Brake at End** to True. Place a second Move Steering block onto the program line with the same parameters, except set **Rotations** to -1. Your program should look like Figure 2-15. Remember to save it before proceeding; that is a good habit to have. Turn on your robot, connect it to the computer, and click the **Download** button to download the program.

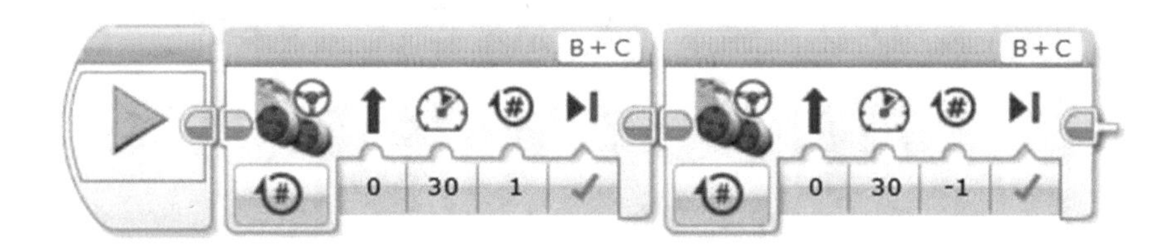

Figure 2-15. *ForwardAndBack program*

This time we will run the program from the brick instead of from the PC. Use the **left** and **right** buttons on the brick to select the second tab on the brick's display, then use the **up** and **down** buttons to highlight the project (MyFirstPrograms), press the **Middle** (dark gray) button to select it, and use the **down** button to highlight the program (ForwardAndBack). The robot is now ready to run. Place it so that it has room to move forward at least one wheel rotation and press the **middle** button to run the program. The robot should move forward one wheel rotation, then backward one wheel rotation, and stop.

Next, let's make this program into a My Block.

Making a My Block

Select the two Move Steering blocks in the program window, but not the Start block. In the menu bar at the top of the page, click **Tools ➤ My Block Builder** to open the My Block Builder window, as shown in Figure 2-16. Type **Fwd_Back** into the Name box. The reason for choosing such a short name is that a short name will be displayed on the block itself, whereas a long name will be truncated. Select one of the icons depicted in the lower half of the window (the Move Steering icon is a reasonable choice). That icon should then appear on the left part of the My Block icon at the top part of the window.

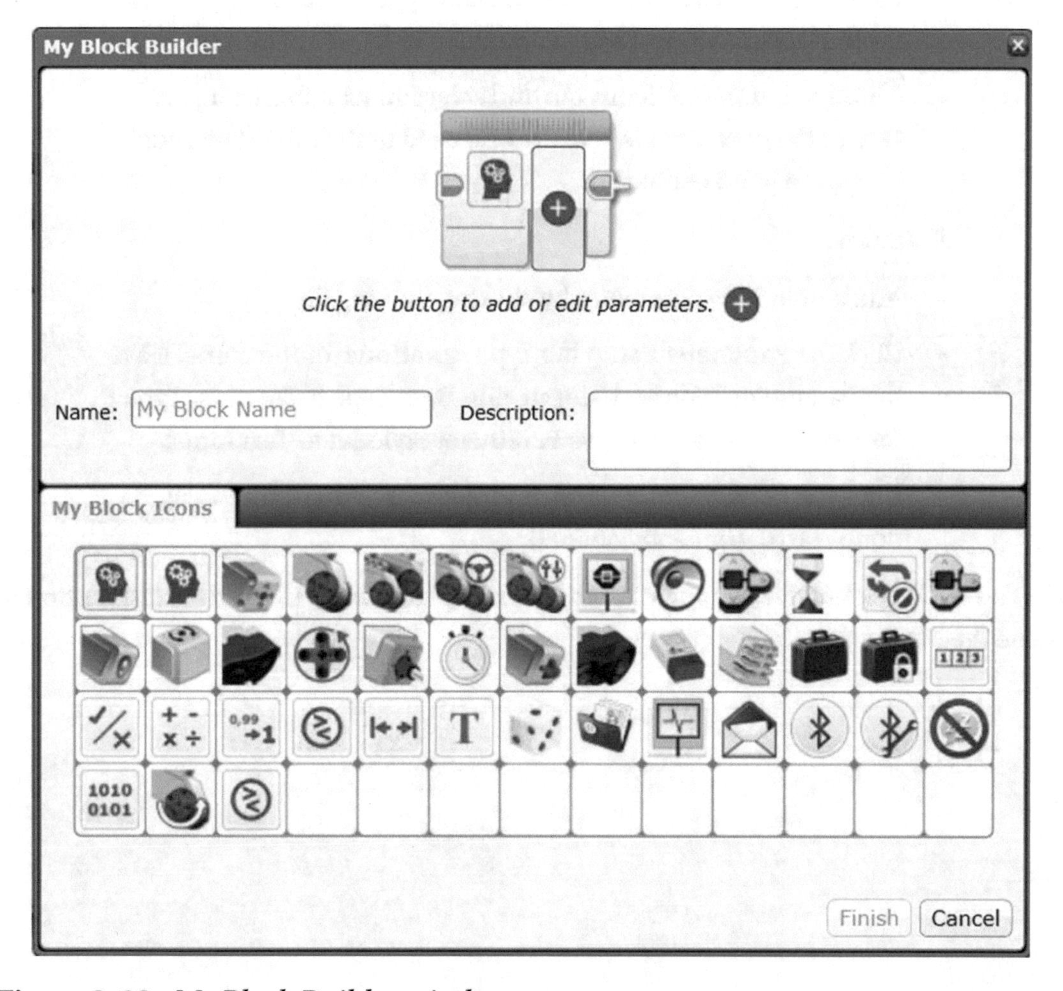

Figure 2-16. *My Block Builder window*

When creating My Blocks, you can set up both input and output parameters. Input parameters are values you set on the My Block, just like the Move Steering block has parameters for Steering, Power, Rotations, and Brake at End (all input parameters in this case). Let's set up this My Block to have input parameters for Power and Rotations.

- **Power**
    - Click the plus sign on the right side of the My Block icon. This should add two new tabs to the lower part of the window: Parameter Setup and Parameter Icons.
    - Click the Parameter Setup tab, type **Power** for the name, leave the Parameter Type set to Input, and Data Type to Number. Type **30** for the Default Value. Leave Parameter Style set to Text Input.
    - Click the Parameter Icons tab and select an icon for the input. One of the power icons, like the one used in the Move Steering block, is a good choice.
- **Rotations**
    - Add a Rotations parameter by clicking the + again.
    - Click the Parameter Setup tab, type **Rotations** for the name, leave the Parameter Type set to Input, and Data Type to Number. Type **1** for the Default Value. Leave Parameter Style set to Text Input.
    - Click the Parameter Icons tab and select an icon (the vertical double arrow is a good choice).

The My Block Builder window should look like Figure 2-17. Click the Finish button to create the My Block.

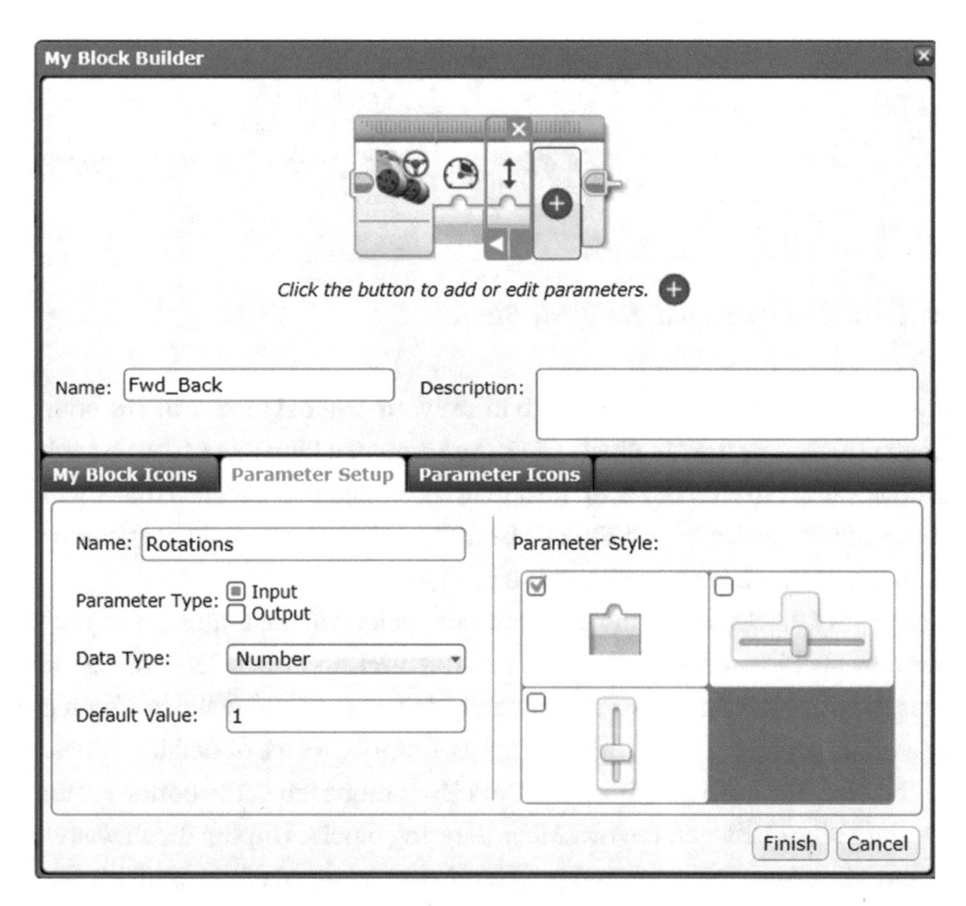

Figure 2-17. *Finished My Block Builder window*

Note that a new tab with the My Block name, Fwd_Back, was added to the program window. Under this new tab, the final step is to connect the inputs from the block at the left side to the appropriate inputs in the two Move Steering blocks. Click the tab below the Rotations input and drag it to the Rotations input of each Move Steering block to make the connection. Do the same with the Power input. Note that you can also drag the data wires to align them for a cleaner look, as shown in Figure 2-18.

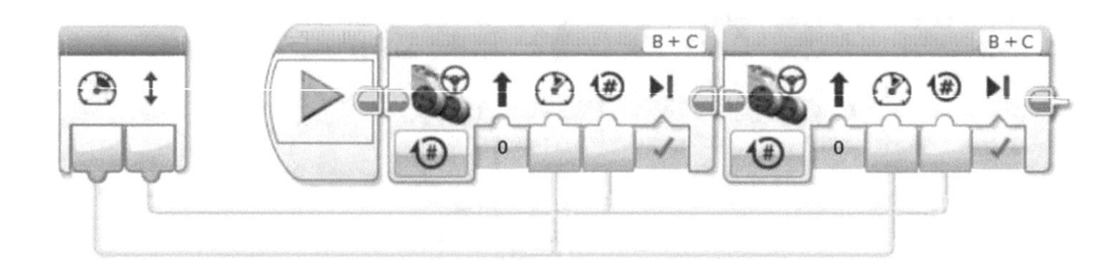

Figure 2-18. *Code for Fwd_Back My Block*

Now, click the **ForwardAndBack** tab to see your original program. The code blocks have been replaced by the My Block. Click the Sequence Plug Exit of the Start block to automatically draw the My Block up next to it. Also, note that because this My Block requires inputs, it cannot be run directly from the brick, even though it shows up on the brick's menu. It must be run from inside a program.

Click the My Blocks tab on the programming palette at the bottom, and you should see your new My Block listed. Download it to the brick and run it. Uh-oh! It goes forward twice instead of forward and back. Fortunately, you can edit a My Block after it is created, as long as you don't need to change the inputs, outputs, icons, or default values. Go back to the Fwd_Back tab and, from the Data Operations tab at the bottom of the page, insert a Math block between the two Move Steering blocks. Unplug the Power wire from the second Move Steering block and plug it into the "a" input of the Math block. Change the mode to multiply, set b = **−1**, and connect the Math block output to the Power input of the second Move Steering block. The corrected code should look like Figure 2-19. Numerous other blocks could be added to the My Block code if desired. If you have not saved the project recently, do so now. Saving frequently is always a good practice.

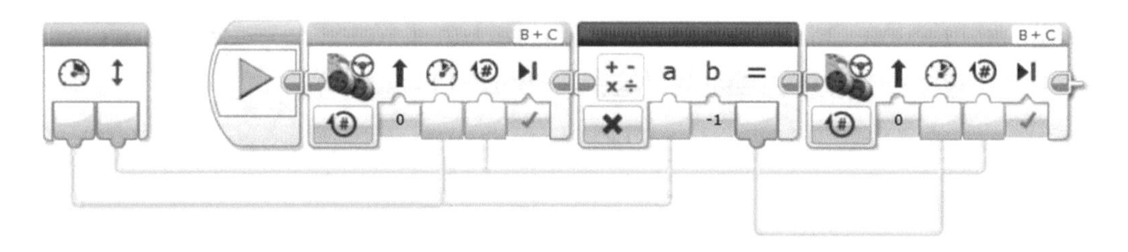

Figure 2-19. *Corrected My Block code*

Working in the Project Properties Window

The last thing we need to do is work with programs and My Blocks in the Project Properties window. Add a new program and name it Temp. Go back to your LightsAndSounds program (or ForwardAndBack), select all of the blocks except the Start block, press **Ctrl+C** to copy them, go to the Temp program tab, and press **Ctrl+V** to paste them into the Temp program. Drag them next to the Start block to connect them to it. The Temp program is superfluous, so let's delete it. Note that if you click the X (on the tab) next to the program name it only closes the program; it does not delete it.

Click the Project Properties tab (the one on the left with the wrench icon). The bottom half of the window has tabs for Programs, Images, Sounds, My Blocks, Variables, and Exportable Items. Your three programs should be listed under the Programs tab (but not Fwd_Back, which is listed under the My Blocks tab). Let's delete Temp. Click Temp.ev3p, then click the **Delete** button to delete it. Careful! The program does not prompt to make sure you really want to delete it. When you click the **Delete** button the program is deleted immediately.

Select LightsAndSounds.ev3p, click the **Copy** button, then the **Paste** button. A copy of the program is added with the suffix 2: LightsAndSounds2.ev3p. It is present, but does not show up on the tabs at the top because it has not yet been opened. You can open it by either double-clicking it in the Project Properties window, or by clicking the Program/Experiment List button and selecting it from the drop-down list. Delete the duplicate program, LightsAndSounds2.ev3p.

Conclusion

At this point you should have written some basic programs, run them on the brick, and implemented a My Block. Now that you are familiar with the EV3 interface and some basic programming techniques, you are ready to begin writing more advanced programs. Chapter 3 starts this process by guiding you to create enhanced My Blocks to make the robot move forward and backward.

CHAPTER 3

Moving Forward and Backward

LEARNING TOPICS

My Blocks, geometry, unit conversion, calibration, feedback, looping constructs, proportional control, variables, conditional constructs, torque, and traction

REQUIREMENTS

1. Move forward or backward a specified distance.
2. Ensure movement is straight (i.e., eliminating the "wiggle").

Simple My Block to Move Forward a Specified Distance

Learning topics covered: My Blocks, geometry, unit conversion, calibration

In virtually every mission a robot attempts, it must move forward or backward, and often both. The Move Steering and Move Tank blocks allow for movement in units of seconds, wheel rotations, or degrees of wheel rotation. Although there could be special circumstances when it would be desired to move your robot for a certain time, normally it is desired to move the robot a certain distance. Thus, the seconds option is typically not useful. Inputting wheel rotations or degrees of wheel rotation is useful, but it is

G. Harding, *Programming LEGO® EV3 My Blocks*, https://doi.org/10.1007/978-1-4842-3438-9_3

handy to translate these units to a linear distance on the mat. In other words, it is much easier to think about telling the robot to go forward 10.5" instead of 1.5 wheel rotations or 540° of wheel rotation.

We can create a My Block to go forward a specified distance in inches by using a Math block and a Move Steering block. We use the Math block, along with a measurement and some simple geometry, to translate inches to degrees of wheel rotation. This process is called *unit conversion*, and it is very common in engineering.

Let's begin by determining the conversion factor to translate inches to degrees of wheel rotation. We need to know the circumference of the tire on your robot. This value can either be measured directly using a flexible measuring tape, or the tire diameter can be measured with a ruler and converted to circumference using the following formula:

$$C = \pi d \tag{1}$$

where C is the tire's circumference and d is the tire's diameter, as shown in Figure 3-1.

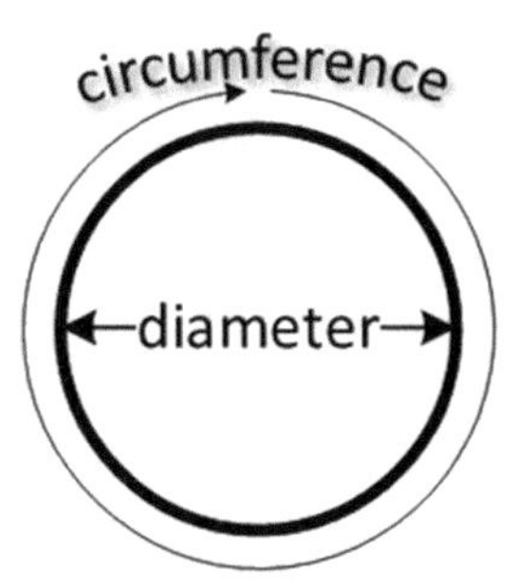

Figure 3-1. *Circle diameter and circumference*

Determine the circumference of your robot's driving tires. Suppose you are using tires that measure 7" in circumference. This is the same as 360° of wheel rotation:

$$7\ inches = 360° \tag{2}$$

Alternatively, we have:

$$\frac{360°}{7\,inches} = \frac{51.4°}{inch} \tag{3}$$

So, if we want the robot to travel 2", we need to tell it to turn the drive motors 102.8°. Notice how the inch units cancel in the following formula:

$$2\,\cancel{inches} \times \frac{51.4°}{\cancel{inch}} = 102.8° \tag{4}$$

Now, let's create a My Block that will use Distance in inches for its input, and convert that number to degrees for a Move Steering block. Open a new program, leaving its default name, Program. Add a Math block to your program followed by a Move Steering block. Set the mode of the Math block to **Multiply**, and the "b" parameter to **51.4**. Set the mode of the Move Steering block to **On for Degrees**. Connect the output of the Math block to the Degrees input of the Move Steering block, as shown in Figure 3-2.

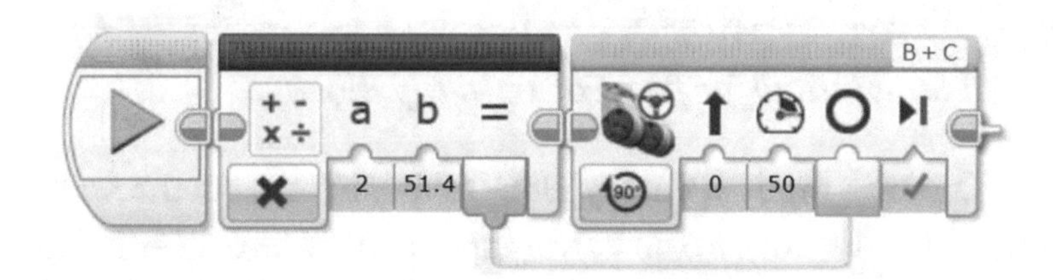

Figure 3-2. *Start of Forward My Block*

Create the My Block by selecting the Math and Move Steering blocks, then go to Tools on the menu bar and select My Block Builder. Name it **Forward** and assign an appropriate icon, such as the Move Steering icon. As we learned in Chapter 2, add a numeric parameter named Power with a default value of 30 and an appropriate icon, and add a numeric parameter named Distance with a default value of 5 (or whatever value you prefer) and an appropriate icon (see Figure 3-3). Once you are happy with the settings, click **Finish** to create your My Block.

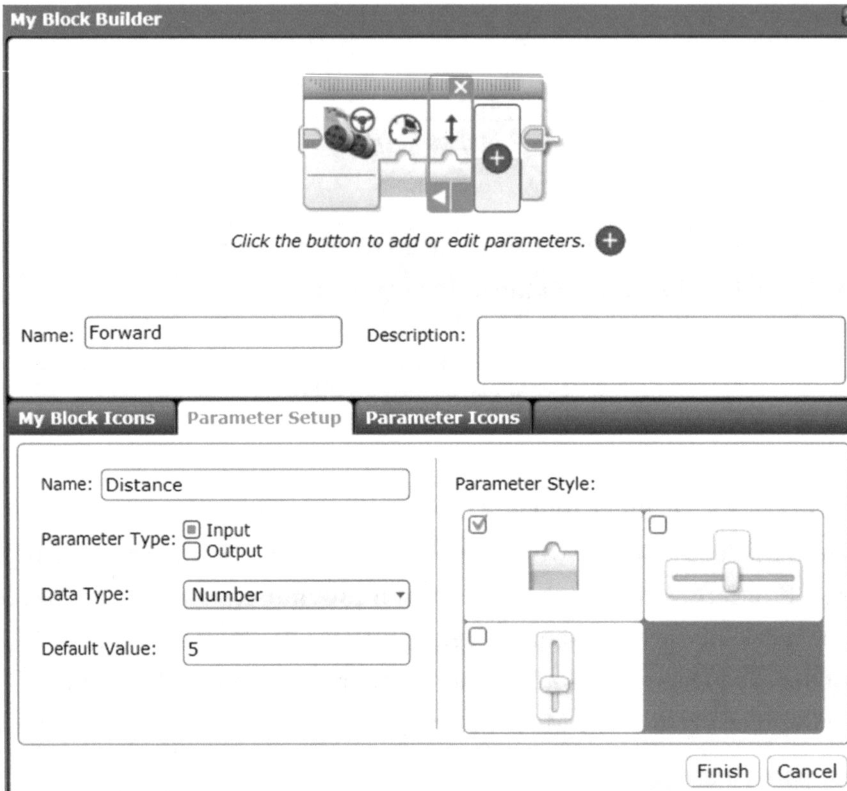

***Figure 3-3.** Builder window for first Forward My Block*

Connect the Distance input of the My Block to the "a" input of the Math block, and the Power input of the My Block to the Power input of the Move Steering block. The programming window showing the Forward My Block should look something like Figure 3-4 (remember to add comments). Move to the Program tab, which should look like Figure 3-5. Try it out on your robot! Remember to run "Program" on the robot, not "Forward." It should go forward about 5" and stop. Experiment with different Power and Distance settings to see how it works.

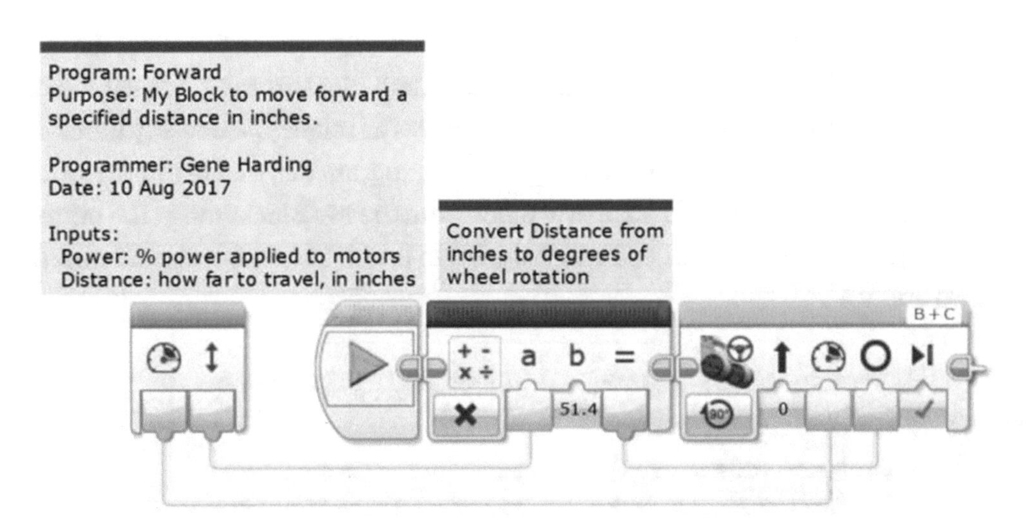

Figure 3-4. *First Forward My Block, complete*

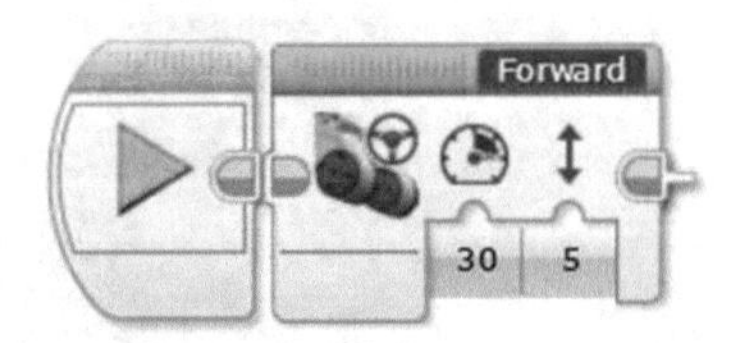

Figure 3-5. *Forward My Block in Program*

There are at least two options for making the robot move backward instead of forward. The first option is to enter a negative number into the My Block for either the Power or the Distance. The advantage of this approach is that a new My Block is not needed; it can be done immediately. A disadvantage is that it can be confusing to understand a program that uses "Forward" My Blocks to move both forward and backward. A second disadvantage is that simply entering a negative value will not work for the advanced version of the Forward My Block that uses acceleration and deceleration to prevent wheel slip (this is covered in Chapter 7).

I recommend creating a second My Block named "Backup," which is very easy to do. Go to the Project Properties tab (far left, wrench icon), then the My Blocks tab within the window, and select Forward.ev3p. Click the Copy button, followed by the Paste button. A new My Block named Forward2.ev3p should now be present. Double-click it to open the new My Block.

Next, double-click the Forward2 tab in the program window, and rename the program Backup. Change the "b" parameter in the Math block from 51.4 to -51.4 so the wheels will rotate in reverse, and adjust the comments to reflect the new My Block's operation, as shown in Figure 3-6. Finally, go to the Program tab, delete the Forward My Block and replace with the Backup My Block from the My Blocks tab at the bottom, save the Project, then download and run Program to test it. Your robot should now move backward about 5".

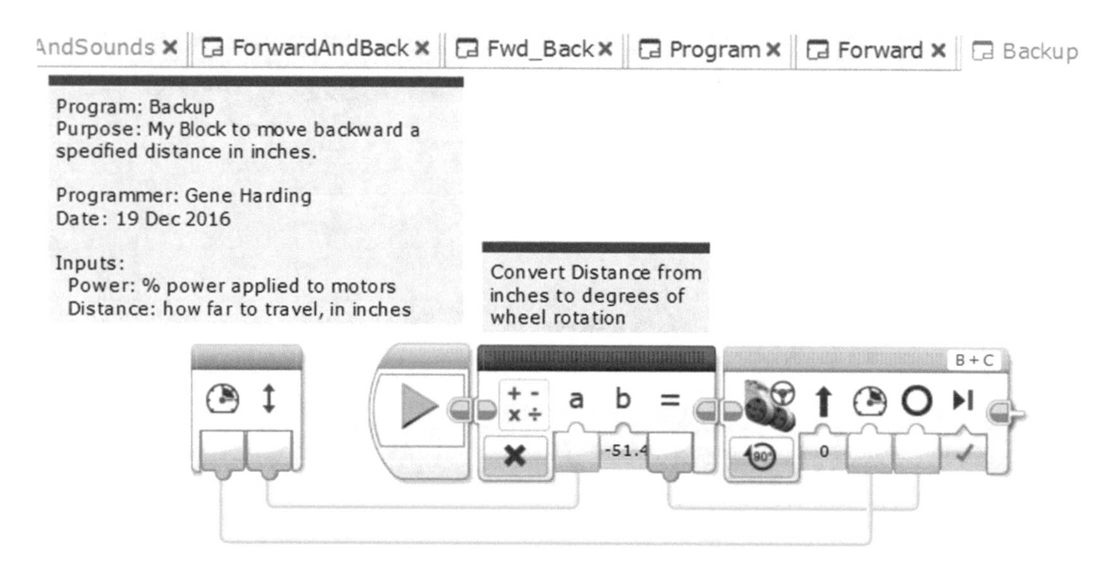

Figure 3-6. *First Backup My Block, complete*

If you want to teach your students about calibration, have them take careful measurements of the robot's forward and backward travel, compare those measurements to the value entered into the My Block, and increase or decrease the conversion factor until the robot travels exactly the distance specified. Note that longer distances are better than shorter distances for this type of calibration, and that low power levels should be used so that wheel slip does not introduce random error into the distance traveled.

Although calibration is a useful concept to teach students, for the purposes of competition it is not practical because most distances traveled on the mat will be determined by trial and error anyway. It is nevertheless helpful to have My Blocks with distance values in inches because it is an intuitive measurement unit.

Eliminating the "Wiggle" Using a Gyro Sensor

Learning topics covered: My Blocks, feedback, looping constructs, proportional control, variables

It is important to "aim" the robot well at the start of a mission set. Aids to aim well include the table walls, markings printed on the mat in the home base area, and a jig custom-built for the purpose. Even if the robot is aimed properly, however, a phenomenon that our team called the "wiggle" can cause problems. Sometimes a robot, although aimed well and programmed to go straight, will steer slightly to the right or left at the beginning of its movement. Even a small such "pointing" error can create a substantial offset from its intended destination if the robot is traveling any significant distance, as shown in Figure 3-7. The problem appears to be caused by mechanical "slop" in the motors. I heard from a tournament judge that one team dealt with the problem by always backing the robot into a wall, with enough power to cause both wheels to slip, before moving forward. This approach seems like a good one to use where possible. There is also a way to detect and quickly correct the problem using a gyro sensor.

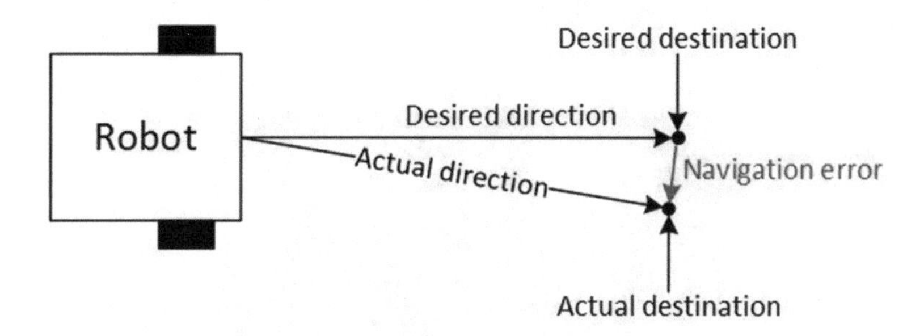

Figure 3-7. *Small "wiggle" can create substantial navigation error*

The basic Forward My Block can be modified to correct the wiggle, but three new concepts are used to do so: feedback, program loops, and proportional control. Feedback will be used from a gyro sensor to determine whether the robot is traveling straight ahead. If the robot veers to the left or right, an error signal will be used to generate steering correction. A program loop will be needed because the process of checking sensor values and correcting the steering must be done rapidly and repeatedly. Implementing this technique creates what is called a proportional controller. A brief overview of this type of controller is in order before making an improved version of the Forward My Block.

Imagine you are driving a car. You look through the windshield to see the road in front of the car. If the car veers toward the left a little bit, you respond by turning the steering wheel slightly to the right so that the car returns to the center of the lane. If the car veers sharply to either side, you respond by turning the steering wheel a large amount in the opposite direction so that the car returns quickly to the center of the lane. The distance of the car from the center of the lane is called the *error*. The center of the lane, or desired value, is called the *set point*. The goal is to stay at the set point (center of the lane) and keep the error at zero. When the error is not zero, the amount of steering adjustment needed is called the *correction*. The term for this approach is *proportional control* because the amount of correction is proportional to the amount of error. Next, we modify the Forward My Block to use proportional control to correct any error caused by the "wiggle."

Begin by attaching a gyro sensor to the robot while the robot is turned off. It can point in any direction as long as the side with the red arrows and dot is facing up, as shown in Figure 3-8. Connect it to one of the sensor ports. It is good practice to not move the robot during the turn-on sequence if it uses a gyro sensor. The gyro sensor tracks the direction in which the robot is pointing, so if the robot veers to one side or the other, the gyro sensor can detect it. It is interesting to note that this is true even if there is wheel slip present.

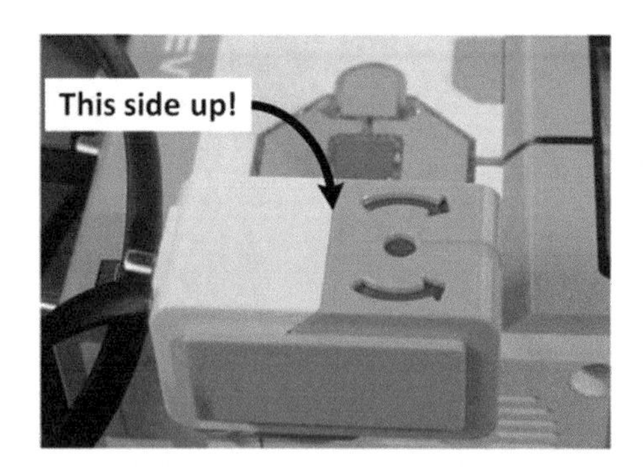

Figure 3-8. *Gyro sensor orientation*

Next, make a copy of the simple Forward My Block using the Copy and Paste buttons in the Project Properties window. Open the new My Block, Forward2, by double-clicking it. While making the changes described in the following paragraphs, it might be helpful to periodically refer to the final program, which is shown in Figure 3-9.

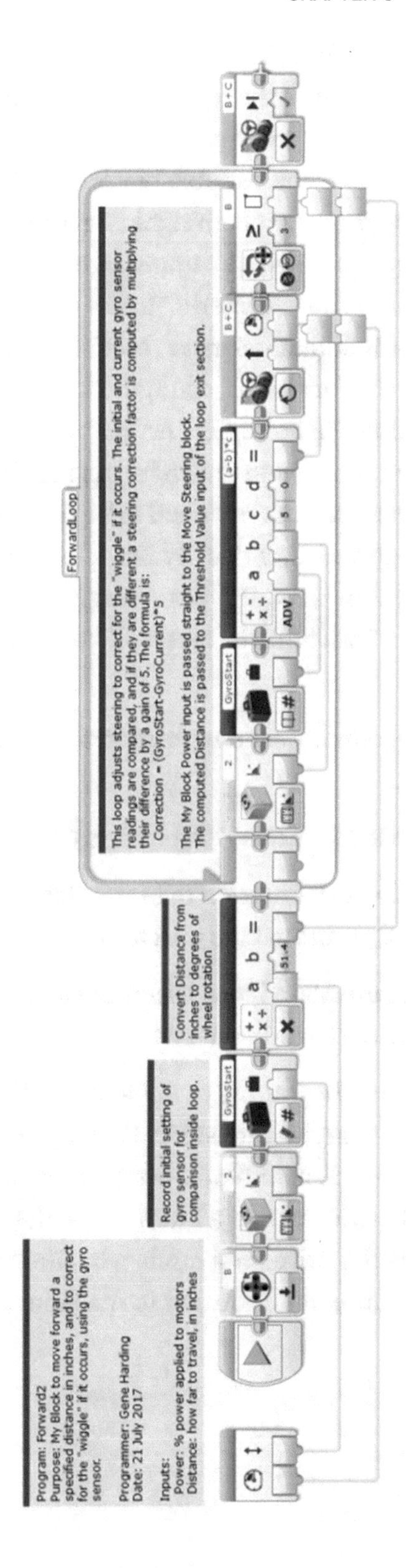

Figure 3-9. *Forward2 My Block to move forward with wiggle correction*

This program needs to read the direction of the robot before any movement. This will be the reference value used to make sure the robot is pointing in the correct direction, and the basis for computing any errors and steering corrections. The element used to store this value is called a *variable.* The Variable block is the first item in the Data Operations tab. Place one on the program chain right after the motor B Rotation Sensor reset block. Click in the white box at the block's upper right, select "Add Variable," name it **GyroStart**, and configure it to "Write - Numeric." Insert a Gyro Sensor block from the Sensor tab, right after the motor B Rotation Sensor reset block, set the port to match the port to which it is connected on the brick, and configure it to "Measure - Angle." Connect its output to GyroStart, so its value will be copied, or "written," to the variable at the beginning of the My Block's operation. Note that, because the gyro sensor's initial value is stored in GyroStart, the gyro sensor does not need to be reset. This preserves its value just in case it is being used before and after this My Block for something else.

In the previous Forward My Block, the Move Steering block determined when the correct distance was traveled. In Forward2 the loop will determine the correct distance, so make the following changes:

- Add a loop to the program flow and reposition the Move Steering block inside the loop.
- Change the mode of the Move Steering block to On.
- Set the mode of the loop exit criterion to Motor Rotation - Degrees, set the condition to 3 (≥), and set the port to B.
- Connect the output of the Math block Distance calculation to the Threshold Value input of the loop exit criterion.

Add a Gyro Sensor block to the beginning of the loop, just before the Move Steering block, set its mode to "Measure - Angle," and set its Port input to the correct port. Next, add a Variable block after the Gyro Sensor block, set its mode to "Read - Numeric," and set its Variable Name to GyroStart. Then add a Math block after the Gyro Sensor block and set its mode to ADV. (Using the Advanced mode will allow for creating a formula with more than two variables.) Connect its output to the Steering input of the Move Steering block.

The Math block will compute the steering correction, but what formula is needed? First, think about how the Steering input works. A negative Steering input to the Move Steering block causes turning to the left; positive input causes the robot to turn right. If the robot veers to the right, a negative Steering input is needed to steer the robot to the left. Turns to the right increase the value of the gyro sensor, so a negative number can be generated by subtracting the sensor value from GyroStart. This value is called the error:

$$Error = GyroStart - GyroSensorValue \tag{5}$$

where GyroStart is the initial value stored at the start of the My Block, and GyroSensorValue is the current value returned by the gyro sensor.

Although the error value could be passed directly to the Steering input of the Move Steering block, in control systems this value is normally multiplied by another number called the gain. If the raw error value is too small, the correction is too slow. In this case a gain > 1 is used. If the raw error value is too large, the system is unstable. Consider the earlier car steering example, and imagine moving the steering wheel a large amount in response to even the smallest error. The result would be that the car would weave back and forth, or even veer off the road: an unstable system. In this case a gain < 1 would be used to stabilize the system.

Multiplying the error by the gain gives the steering correction:

$$Correction = Error * Gain \tag{6}$$

$$Correction = (GyroStart - GyroSensorValue) * Gain \tag{7}$$

Do the following to implement the correction formula:

- Connect the GyroStart variable output within the loop to the “a” input of the Math block.
- Connect the Gyro Sensor block output within the loop to the “b” input of the Math block.
- Set the “c” input to **1**.
- Click in the formula box at the upper right and type the formula **(a-b)*c**.

At this point the data wire connecting to the Threshold Value of the Loop block is probably misaligned because of the extra blocks added to the loop. To fix this, unplug the data wire from the loop, let it go, and redo the connection. Next, connect the My Block Power input to the Power input of the Move Steering block (same as with the Forward My Block).

The last thing to do before tuning is to make sure the robot stops as soon as it reaches the specified distance. Add a Move Steering block to the end of the program chain, after the Loop block. Set its mode to Off and Brake at End parameter to True (check mark). The My Block is now ready to be tuned (i.e., figure out what value to use for the Gain parameter).

It is often effective to begin tuning by setting the Gain parameter to 1, then running the program. If it oscillates back and forth instead of moving forward, the system is unstable and the gain needs to be reduced. Reduce the gain by an order of magnitude (i.e., from 1 to 0.1, then to 0.01, etc.) until it does not oscillate. If the system does not oscillate, increase the Gain by an order of magnitude (1 to 10, then to 100, etc.) and try it again. Continue increasing the Gain until it oscillates, then reduce the Gain until it is stable. Oscillation is bad, but a Gain value near the point where it starts to oscillate is good because the steering is corrected faster. The appropriate Gain value will depend on characteristics of the robot, including wheel base and tire diameter. For the robot used to test these programs, a final value of Gain = 5 worked well. The final program is shown in Figure 3-9.

Forward2 can easily be copied and modified to create a Backup2 My Block that works the same way in reverse, but four changes must be made after copying the program.

- The Distance conversion factor must be negated because the motor counters will be starting at zero and decrementing to negative values.
- For the same reason, the Compare Type to exit the Loop block must be changed from "≥" to "≤."
- The correction formula must be reversed. This can be done by changing the formula from "(a-b)*c" to "(b-a)*c," or by negating the gain value, "c." (Make one of the two changes, but not both. Doing both will result in the same formula because the two "negatives" will cancel each other.)
- A Math block must be added before the loop to negate the Power input before it is passed to the Move Steering block inside the loop.

Remember to adjust the comments appropriately, as well. Backup2 should look similar to Figure 3-10.

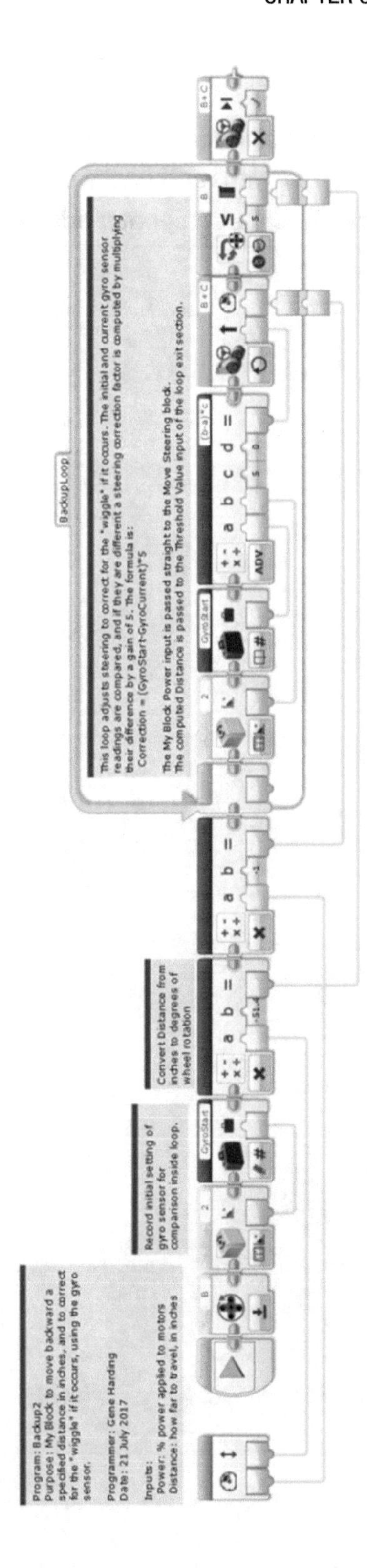

Figure 3-10. *Backup2 My Block*

Conclusion

This chapter covered how to make your robot move forward and backward in a straight line for a specific distance. Although forward and backward movements are important for navigating, they are not sufficient. The next chapter addresses turning.

CHAPTER 4

Turning in Place

LEARNING TOPICS

My Blocks, geometry, unit conversion, calibration, torque, traction, dead reckoning, and parallel execution

REQUIREMENT

Turn left or right to an angle specified in degrees

The next step in effective navigation is being able to turn. This chapter focuses on turning in place from a dead stop. There are two basic approaches. The first is to lock one wheel while driving the other wheel. In this mode, the robot's pivot axis is the wheel that is not rotating, as shown in Figure 4-1 (a–d). The second approach is to drive both wheels, but in opposite directions. In this method, the robot's pivot axis is ideally the center point between the two wheels, as shown in Figure 4-1 (e and f). For the sake of clarity, the term "turn" is used to describe the first technique, whereas "spin" is used to describe the second.

G. Harding, *Programming LEGO® EV3 My Blocks*, https://doi.org/10.1007/978-1-4842-3438-9_4

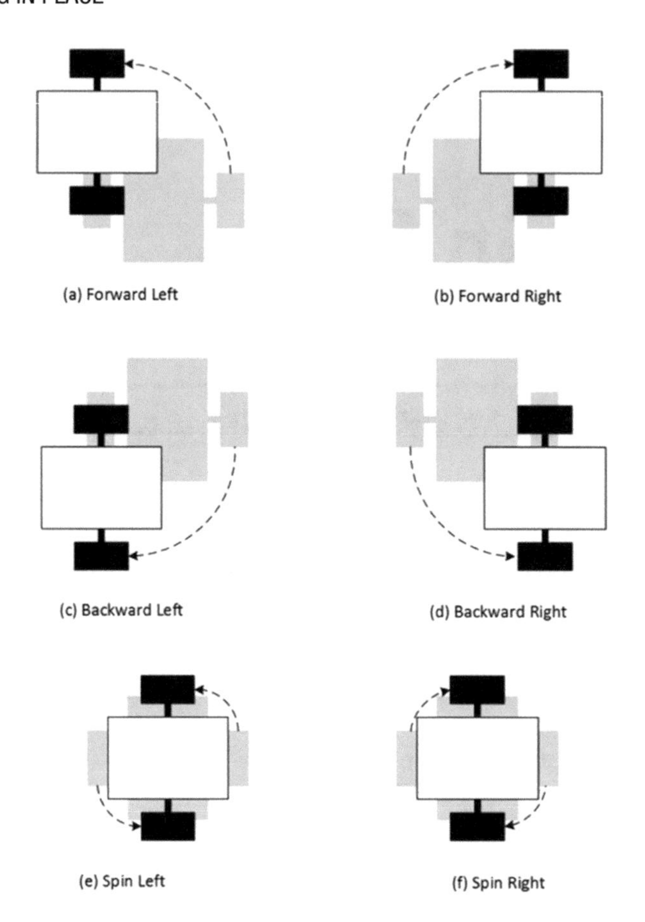

Figure 4-1. *Turning options*

For high turning precision, especially for the first approach previously listed, a narrow, high-crown tire is better than a wide tire with a flat cross-section. As shown in Figure 4-2, the reason is that the pivot point for a high-crown tire is the center of the tire, whereas the pivot point for a flat-profile tire could be most any point across the part of the tire contacting the mat.

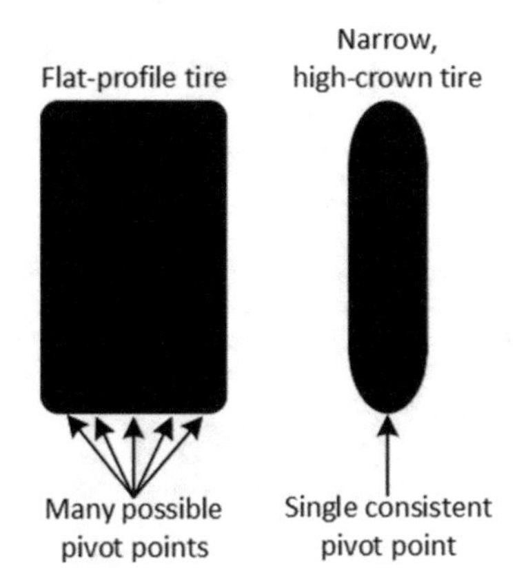

Figure 4-2. *Tire profile comparison*

Turning Left and Right

Learning topics covered: My Blocks, geometry, unit conversion, torque, traction

Although it is possible to construct one My Block that can perform all four turn types (a–d), it is arguably better to use one My Block for right turns and another for left turns. There are at least two reasons to do so. First, the code becomes somewhat self-commenting as a block named TurnRight clearly implies the robot is going to turn right, even with no comment added. Second, a fringe benefit is that the My Block code is simpler.

To create an intuitive turn My Block, desired degrees of turn angle for the robot must be converted to degrees of wheel rotation for the appropriate drive motor. In the forward and backup My Blocks, we converted inches of linear movement to degrees of wheel rotation. A similar exercise in geometry and algebra will allow a straightforward conversion from turning angle degrees.

The distance (center to center) between the robot's two drive wheels is the track width. For turning, this distance is also the radius of the circle that the outside wheel traces as the robot turns, as shown in Figure 4-3. The lowercase variable c refers to the circumference of the driving wheel, R is the robot's track width and turning circle radius, and C is the robot's turning circle circumference. Dividing the turning circle circumference by the wheel circumference gives the ratio of motor (axle) rotation degrees to turn angle degrees, which is the number we need to convert a turning angle My Block input to motor rotations. If the robot track width is 4.8" and tire circumference is 7.0", then we get the following:

$$conversion\ factor = \frac{°motor\ rotation}{°turn\ angle} = \frac{2\pi R}{c} = \frac{2\pi(4.8")}{7.0"} = 4.3 \tag{8}$$

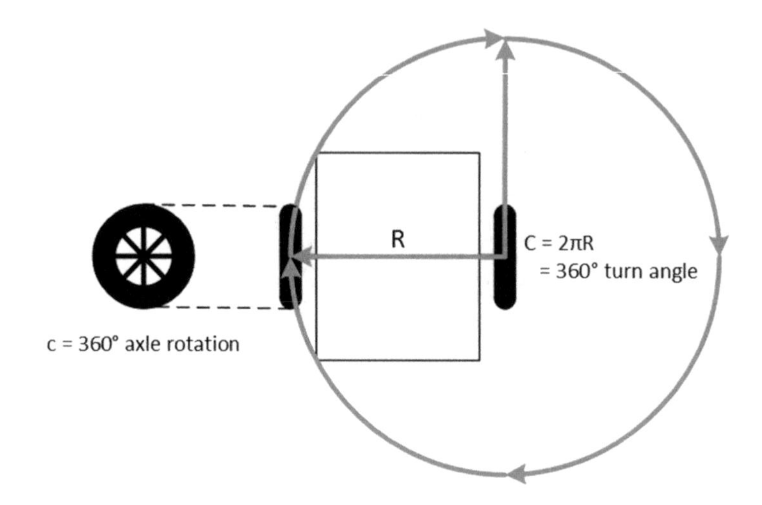

***Figure 4-3.** Degrees of turning angle vs. axle rotation*

The My Block is pretty simple. Start with a Large Motor block configured to stop motor C so that it does not move, then a Math block to convert turning angle to degrees of motor rotation, and finally a Large Motor block to rotate the B motor (left side) to make the robot turn to the right. The TurnRight My Block code is shown in Figure 4-4.

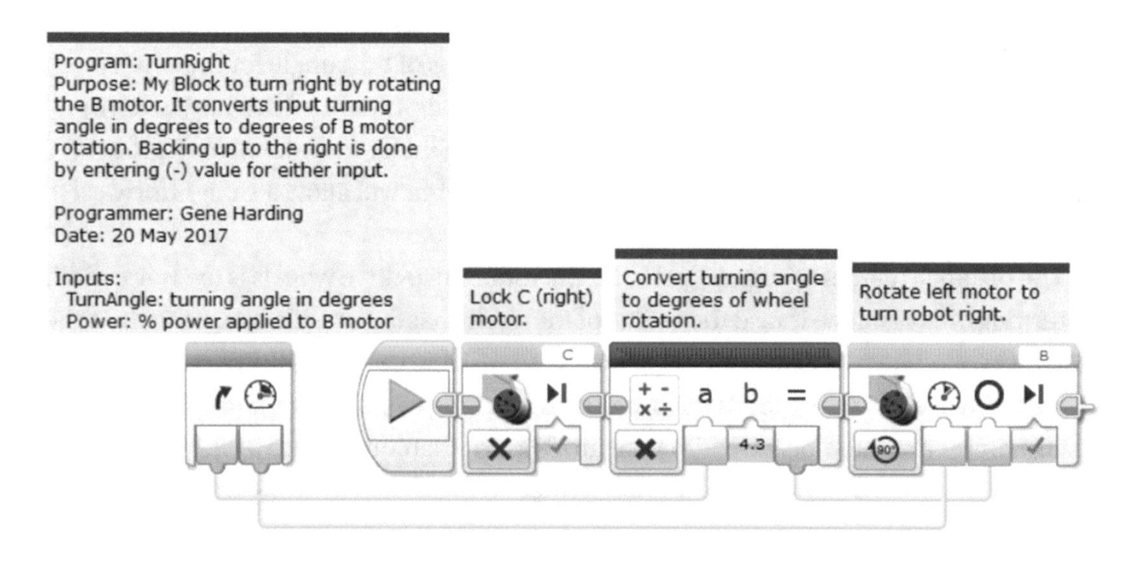

***Figure 4-4.** TurnRight My Block code*

The My Block itself is shown in a program in Figure 4-5, where it has been set to turn the robot 20° at 20 percent power. The My Block code (Figure 4-4) will multiply the 20° input by 4.3 so the wheel rotates 86° to accomplish the 20° turn angle. Be careful not to use a power that is too large. This can cause the drive wheel to lose traction, and the resultant slippage will make turning angles inconsistent. A little bit of testing and tweaking to calibrate the conversion parameter and input power should give you a My Block that turns at angles close to the specified input values. Also, note that a negative value for either input will make the robot back up to the right, whereas positive values will make it rotate forward to the right.

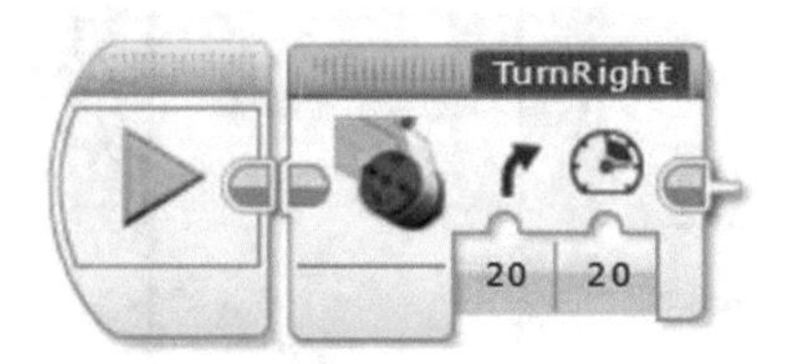

Figure 4-5. *TurnRight My Block in a program, to turn 20° at 20 percent power*

A TurnLeft My Block is now easy to make. Use the same approach, but change the TurnAngle icon from a right arrow to a left arrow, change the first motor block to lock the B motor, change the second motor block to drive the C motor, and update the comments appropriately. Figure 4-6 shows an example.

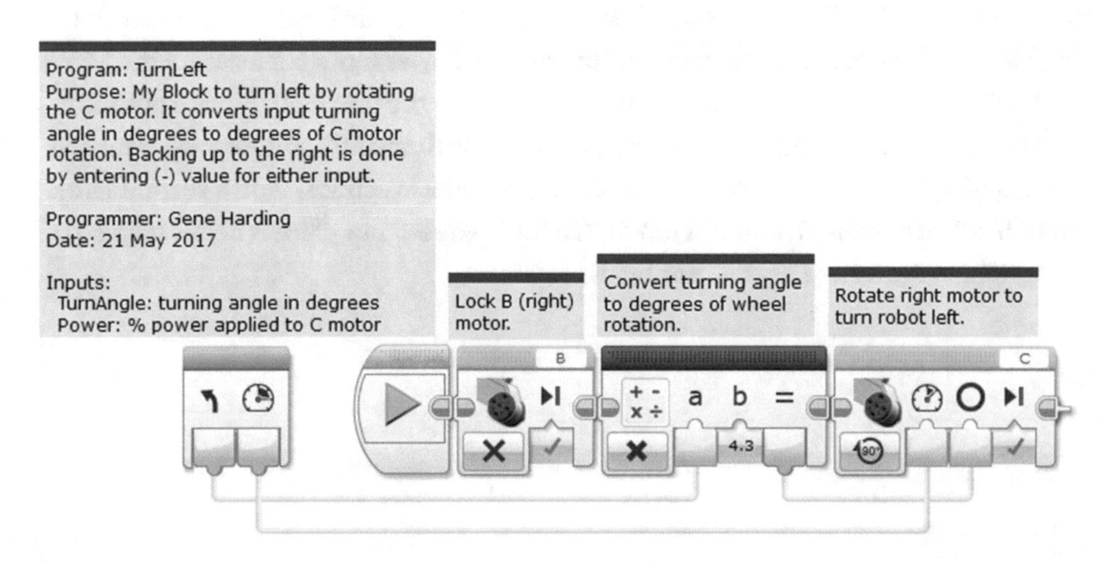

Figure 4-6. *TurnLeft My Block*

Spinning Left and Right

Learning topics covered: My Blocks, geometry, unit conversion, torque, traction

The SpinRight and SpinLeft My Blocks are similar to the turn My Blocks, but the geometry is different because both motors are driven together. In the turn My Blocks, the robot's track width was the turning circle radius. In the spin My Blocks, the track width is the turning circle diameter (see Figure 4-7).

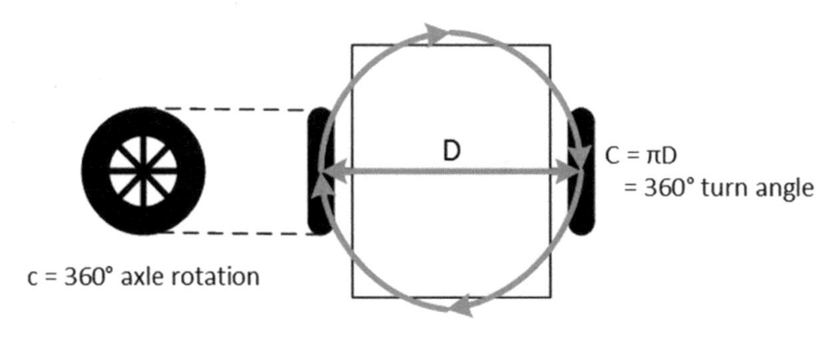

Figure 4-7. *Degrees of spin angle vs. axle rotation*

This means that the conversion factor for the spin My Blocks will be about half of the conversion factor in the turn My Blocks. Because both motors are driven, the first block will be the Math block to convert turning angle into degrees of wheel rotation, with the smaller multiplier. Add a second Math block to invert (multiply by -1) the output for the right motor, which will turn backward. As in the TurnRight My Block, next comes the Large Motor block to rotate the B motor forward. Its Power input comes from the My Block's Power input, and its Degrees input comes from the first Math block's output.

Now we will do something that is a new feature with the EV3 software: operate two sections of code in parallel simultaneously, i.e., parallel execution. Add a second Large Motor block to run the C motor in mode "On for Degrees," but place it below instead of to the right of the other Large Motor block.

Next, use the pointer to "grab" the Sequence Plug Exit of the second Math block (not the data wire output port; the output plug at the right end of the Math block) and connect it to the Sequence Plug Entry of the C motor block. When you do that, the program will automatically shift the other motor block to the right, making it obvious that parallel operation is occurring. Connect the numeric data output of the second Math block to the Degrees input, and the Power input of the My Block to the motor block's input, as shown in Figure 4-8.

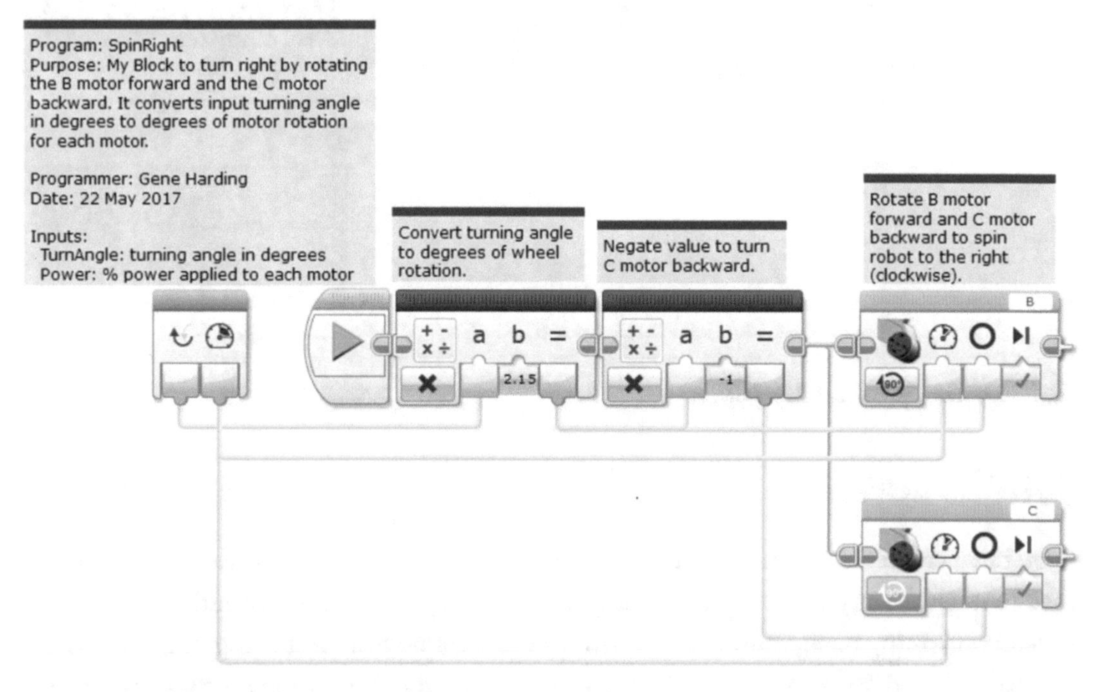

Figure 4-8. *SpinRight My Block*

Construct the SpinLeft My Block in a similar manner, changing the forward drive motor to C, the backward drive motor to B, and updating the icon and comments, as shown in Figure 4-9.

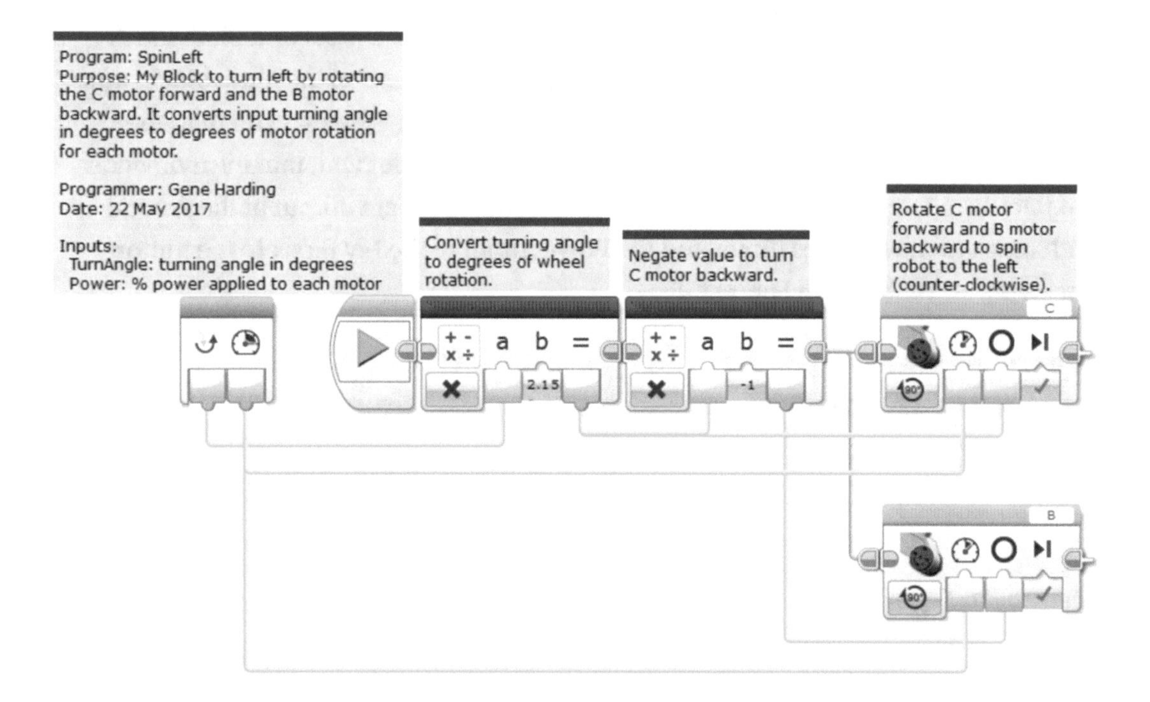

Figure 4-9. *SpinLeft My Block*

Conclusion

If you have created all of the My Blocks up to this point in the book, you have what you need to navigate anywhere on a challenge board using what is called dead reckoning. In dead reckoning navigation, one's position is calculated from a starting point based on the distance and direction traveled. The problem with dead reckoning is that any errors in those calculations tend to be larger as distance traveled gets larger. For instance, if you point your robot at an object on the board and program it to go forward, it might come within a fraction of an inch of its intended mark if that mark is a short distance away. On the other hand, if it must travel several feet to get there, it could be off by more than an inch, which could result in a failed mission.

Before the days of radio navigation (e.g., LORAN, TACAN, GPS), pilots used what is called waypoint navigation. People use this method of navigation all the time without thinking about it. Suppose you were giving someone directions to the store. It might go something like this:

> *"Go up the street that direction and take a left at the stop sign. Turn right at the first traffic light, then go about ¾ mile and turn left just before the gas station. The store will be on the right."*

Those landmarks—stop sign, traffic light, gas station—are waypoints, fixed points of reference that can be used to adjust, or correct, dead reckoning navigation. FLL challenge boards always have lines on the mats and border walls that can be used for such corrections to improve navigation precision. The next two chapters deal with using lines and border walls for this purpose.

CHAPTER 5

Using Lines

LEARNING TOPICS

My Blocks, unit conversion, feedback, looping constructs, proportional control, variables, and conditional constructs

REQUIREMENTS

1. Find a black line when crossing it.
2. Follow the right or left edge of a black line a specified distance.
3. Find the point at which a black line makes an abrupt 90° change in direction.
4. Square up on a black line.

The black lines on FLL challenge mats can be used to precisely locate the robot in one direction (north–south or east–west), or in the case of lines intersecting at 90° angles, provide a precise waypoint in both directions (north–south and east–west). Figure 5-1 shows the three common line configurations seen on most competition mats: lines, line-Ts, and line-Ls. The EV3 color sensor can be used to detect and follow the high-contrast edge where a black line transitions to a white line.

G. Harding, *Programming LEGO® EV3 My Blocks*, https://doi.org/10.1007/978-1-4842-3438-9_5

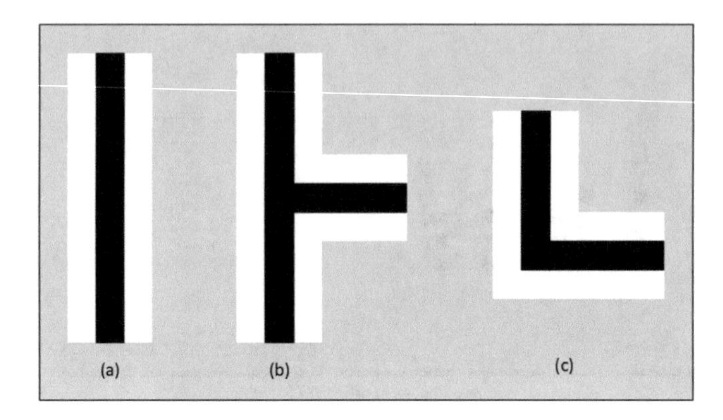

Figure 5-1. *Line, line-T, and line-L*

The color sensor has several different modes, most of which deal with detecting either a color, reflected light intensity, or ambient light intensity. Color sensing is covered here for completeness, but the focus in this chapter is dealing with black and white lines.

When used in a color-sensing mode the sensor can measure the following colors: no color (0), black (1), blue (2), green (3), yellow (4), red (5), white (6), or brown (7), outputting the numeric value listed in parentheses corresponding to the detected color. It can also compare the numeric value of the detected color to a given value, outputting both a logic result of the comparison and a numeric result for the color detected.

The color sensor can measure either reflected light intensity or ambient light intensity, outputting numeric values ranging from 0 (minimum reflected light) to 100 (maximum reflected light). When measuring reflected light intensity, the red light-emitting diode (LED) inside the sensor shines light downward, and the sensor detects how much light is reflected back. When measuring ambient light intensity, the LED remains off, so only ambient light is detected (i.e., light from other sources, such as lamps or overhead lights in the room or sun shining through the window). The sensor can also compare either reflected light intensity or ambient light intensity. In these modes it still outputs the numeric value of light intensity measurement, but also outputs a logic value corresponding to the result of the comparison.

In each of the Compare and Measure modes just listed, the sensor's sample rate is 1 kHz, or 1,000 samples per second (EV3 User Guide, either Home or Education Edition).

Finally, the color sensor has Calibrate modes. If a guard is built around the sensor to prevent ambient light from entering, the Calibrate modes might not be needed. On the other hand, if a guard is not in place to shield it and only one sensor is used for line following, calibrating the sensor is very important. The next section addresses calibrating the color sensor for different ambient light conditions.

Calibrating the Color Sensor

Learning topics covered: calibration, feedback, looping constructs

In Reflected Light Intensity mode, the EV3 color sensor can return values ranging from 0 to 100. A 0 value indicates no reflected light, corresponding to a surface so dark that none of the LED's light is reflected into the sensor. A value of 100 indicates "full" reflection, or a surface so bright that it reflects essentially all of the LED's light into the sensor. In most situations, the range of sensor values will not be 0 to 100, but instead a smaller range. For example, suppose that a given competition mat with a certain amount of ambient lighting results in a light intensity value of 9 over a black line (dark surface) and a 93 over a white line (reflective surface). The problem is that different levels of ambient lighting can change these maximum and minimum values, so a given program can operate differently in different environments. My teams have seen their programs operate differently as the sun changed position in the sky, shining more or less light into the room through the windows.

Calibration can deal with this problem by redefining what constitutes a value of 0 or 100. For instance, in the scenario described earlier, the robot could be calibrated to treat a 9 as a 0 value, and a 93 as 100. By recalibrating the robot whenever the environment changes, you can reduce or eliminate the impact of a change in ambient light conditions. My teams found calibration to be critical when using a robot with no guard, or shielding, around the light sensor(s), so we always calibrated the robot just before doing a formal mission set run. We later built a guard to go around the light sensors to block ambient light. Although the guard seemed to eliminate the need for regular calibration, we continued the practice just to be safe.

There are three Calibrate modes, all dealing with reflected light intensity. The Minimum mode redefines what constitutes a 0 value; the Maximum mode redefines what constitutes a value of 100; and the Reset mode changes back to the default values. Let's start with a basic calibration program.

The procedure to calibrate reflected light intensity is as follows:

1. Reset the sensor.
2. Place the robot so its light sensor is over a black line.
3. Read the sensor value and pass it to the Calibrate Minimum block.
4. Place the robot so its light sensor is over a white line.
5. Read the sensor value and pass it to the Calibrate Maximum block.

Start the program with a Color Sensor block in "Calibrate - Reflected Light Intensity - Reset" mode. Add a Wait block in "Brick Buttons - Compare" mode to wait for a button to be bumped. This allows for the user to place the robot over a black line before doing the minimum intensity calibration. Next, add a Color Sensor block in "Measure - Reflected Light Intensity" mode to read the value, and connect it with a data wire to a Color Sensor block in "Calibrate - Reflected Light Intensity - Minimum" mode to do the calibration. Add another Wait block in "Brick Buttons - Compare" mode to wait for a button to be bumped. This allows for the user to place the robot over a white line before doing the maximum intensity calibration. Next, add a Color Sensor block in "Measure - Reflected Light Intensity" mode to read the value, and connect it with a data wire to a Color Sensor block in "Calibrate - Reflected Light Intensity - Maximum" mode to do the calibration. The program could be named Calibrate, and should look like Figure 5-2. (Note that this is just a program, not a My Block.)

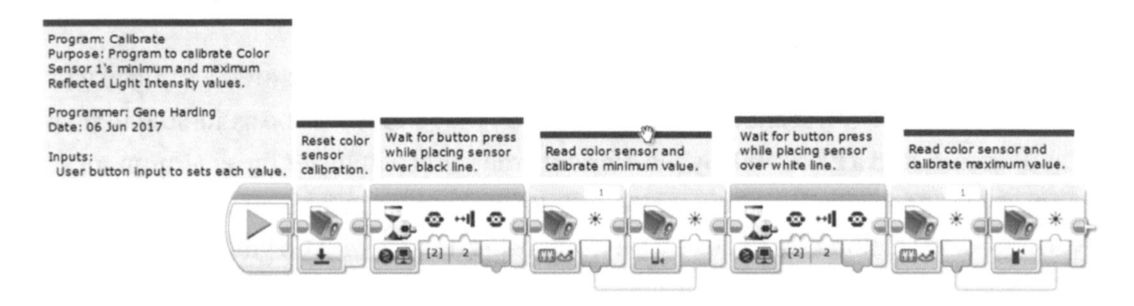

Figure 5-2. *Calibrate program*

This approach works, but there is a significant issue with it: There is no way of being sure that the light sensor is placed so that it reads the desired minimum or maximum value. We address that issue by using feedback and a couple of loops.

Copy Calibrate to a new program and name it Calibrate2. We will use a loop to continuously poll the color sensor and display its value so we know when it is at a minimum or maximum. Replace the Wait blocks with Loop blocks. Set each loop exit mode to Brick Buttons, and pick the button of your choice. Note that when using the buttons in this manner, using the "bumped" option (2) is probably the most reliable choice. Sometimes one press or release can trigger the exit for multiple loops, even though it should not do so. Give a meaningful name to each loop, such as DisplayMin for the first loop and DisplayMax for the second. Place a Color Sensor block followed by two Display blocks (from the Action Blocks tab) inside each loop. Configure both Color

Sensor blocks in "Measure - Reflected Light Intensity" mode. The Display blocks will present a user prompt and display the current light intensity value.

Display blocks have several different modes, but the one we need now is "Text - Grid." According to the EV3 help, this mode divides the brick's display into a rectangular grid of 12 rows by 22 columns numbered 0 to 11 and 0 to 21, respectively. Each row is 10 pixels high, and each column is 8 pixels wide. Click on the small window at the upper right of the block where it says "MINDSTORMS." This is the text to be displayed. Replace it with "Min?." Set the Clear Screen parameter to **True**. This clears the entire screen each time the loop executes. Because this is the first item to be displayed, set the Row and Column parameters both to **0**. The second Display block's mode will be the same, but instead of typing a text value in the display box, select the "Wired" option. Notice that this generates a new input called Text in the Display block. Connect the Color Sensor output to this input. Now the program will display the sensor value each time it updates the minimum calibration. For the second Display block, set the Clear Screen parameter to **False** so it does not erase the "Min?" prompt. Set the Column to **12** or so and leave the Row set to **0** so the light intensity value is displayed on the same line, but after the "Min?" prompt. See Figure 5-3 for an illustration.

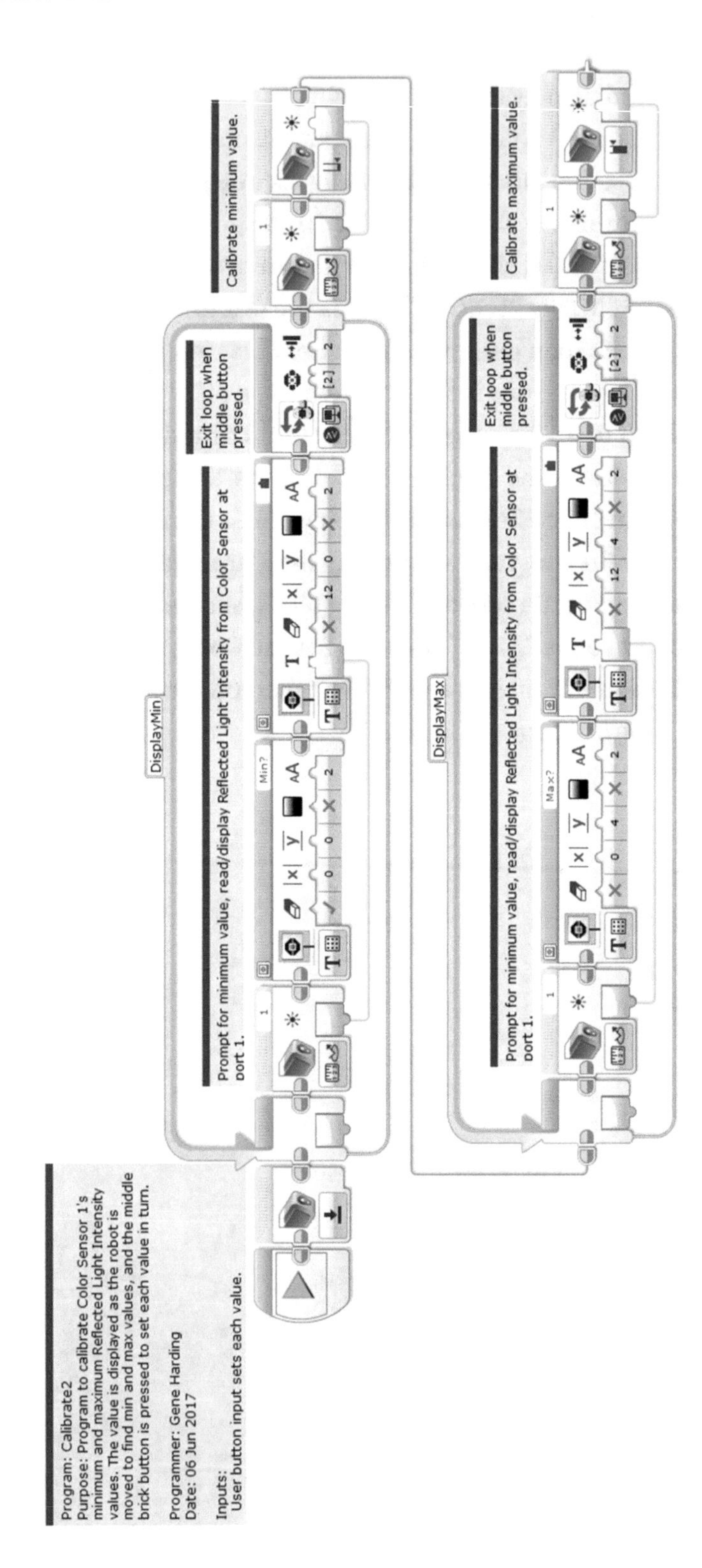

Figure 5-3. *Calibrate2 program*

The second loop will be similar, but not the same as the first. To leave the Minimum value displayed, set the Clear Screen parameter of both Display blocks to **False**. Set the prompt text to "Max?," Column to **0**, and Row to **4** so it will display at the start of a new line. As with the DisplayMin loop, select the "Wired" option and wire the output of the Color Sensor block to the Text input of the second Display block. Set the Column parameter to **12** and Row to **4**.

One tip here is very important: Add several spaces to the end of the "Max?" prompt so that the numeric value is cleared each time the loop executes. This is important to avoid confusing readouts. For instance, suppose the sensor is placed over a white area and returns a value of 100, then is moved to a shaded area where it returns a value of 25. If the old value is not cleared (with the extra spaces), the 2 will overwrite the 1 digit of 100, the 5 will overwrite the first 0 digit of 100, and the second 0 digit of 100 will remain, so it will look like the value being displayed is 250 instead of 25.

Note in Figure 5-3 that the program chain has been split into two rows. Arranging programs in this manner often improves readability, so let's try it. Select the second loop and the blocks that follow it, and drag them down and to the left so they are under the first loop. Then, "grab" the Sequence Plug Exit of the top right Color Sensor block and connect it to the Sequence Plug Entry of the second loop. Remember to add appropriate comments, then save and try out the Calibrate program.

Finding a Black Line

Learning topics covered: My Blocks, feedback, looping constructs

Before the robot can follow a line, it must first find the line. Although this process is fairly straightforward, some patterns used on the competition mats make it a bit more complicated. It is now quite common for mats to have patterns (separate from the lines) with dark areas that look black (or are black) and light areas that look white (or are white). Because of this, my teams generally found it effective to use the following steps to find a black line on the mat:

1. Use dead reckoning to get close to the line.
2. Look for the white space before the black line.
3. Look for the black line itself.

The first step bypasses most or all of the areas on the mat that might be mistaken for the black line or its surrounding white space. The second step is especially useful when there is a dark pattern near the line, and the dark pattern could be mistaken for the black line itself. Figure 5-4 shows a black line with three patterns nearby that might appear black to a robot's light sensor. By looking for the white line first, the robot will bypass the star, circle, or chevron if any of those is in its path. Using dead reckoning followed by the white space search greatly increases the probability of finding the desired black line.

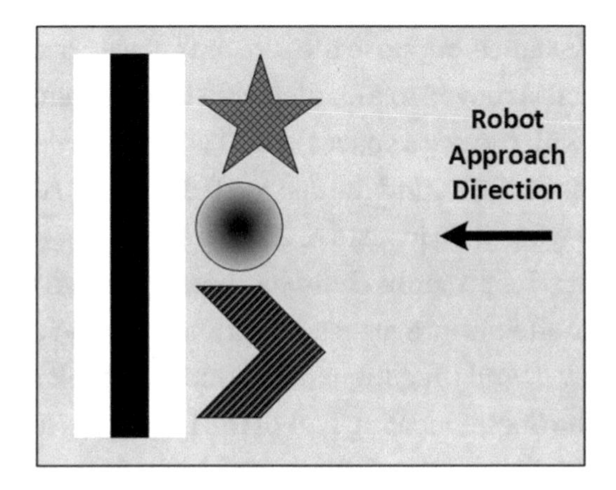

Figure 5-4. *Dark areas adjacent to black line*

Dead reckoning navigation was covered in the two preceding chapters. A color sensor in Reflected Light Intensity mode can detect the black line and its surrounding white lines. Let's begin by making a program to find the white area, starting with the loop used in the earlier forward My Blocks. Copy the "wiggle-correction" loop from Forward2 and paste it into a new program. Because the goal is not to move a specified distance, that part of the code is not needed, and the loop exit criterion must be modified. Change the Loop block mode from "Motor Rotation - Compare - Degrees" to "Color Sensor - Compare - Reflected Light Intensity" and make sure the Port is set to "1". A high Threshold Value is needed to detect white, probably somewhere between **85** and **95** because a white surface should reflect almost all of the light, and a properly calibrated sensor should return a value close to 100. Name the loop something meaningful like **FindWhite**. Add a Move Steering block after the loop to stop both motors, and test the program. If a mat is not available, place a white sheet of paper on a relatively dark surface, such as a woodgrain table, and verify that the robot stops when it reaches the paper. The program should look something like Figure 5-5.

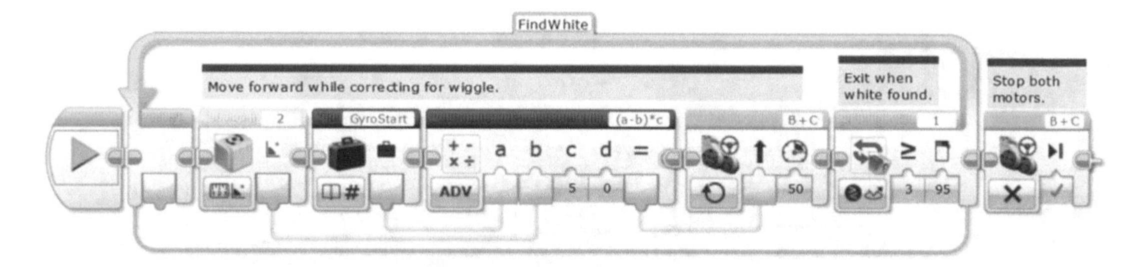

Figure 5-5. *Program to find white area*

A second loop is needed to find the black line, so copy the loop FindWhite, paste it right after the first one, and rename it **FindBlack**. Because the goal is to find black, which reflects very little light, the loop exit criterion will be to a light intensity less than a low value, probably somewhere between **5** and **15**. Change the Compare Type parameter for the loop to **5** (≤) and enter a low threshold value. Add a black line to the white piece of paper and test the program.

Because this program might be needed many times across multiple mission sets, it makes sense to convert it to a My Block, so select everything but the Start block and open the My Block Builder. Reasonable numeric inputs could include minimum threshold with a default value of 5, maximum threshold with a default value of 95, power for the motors with a default of 20, and port number with a default of 1. Although the reason might not be apparent until later in this chapter, it is often useful to be able to specify the port number when using sensors. Click the Port window at the upper right of each Loop block and select the Wired option, then connect the PortNumber input of the My Block to the appropriate input of the Loop block. The final My Block should look something like Figure 5-6. Test it to make sure it works inside a program and remember to add comments before moving on.

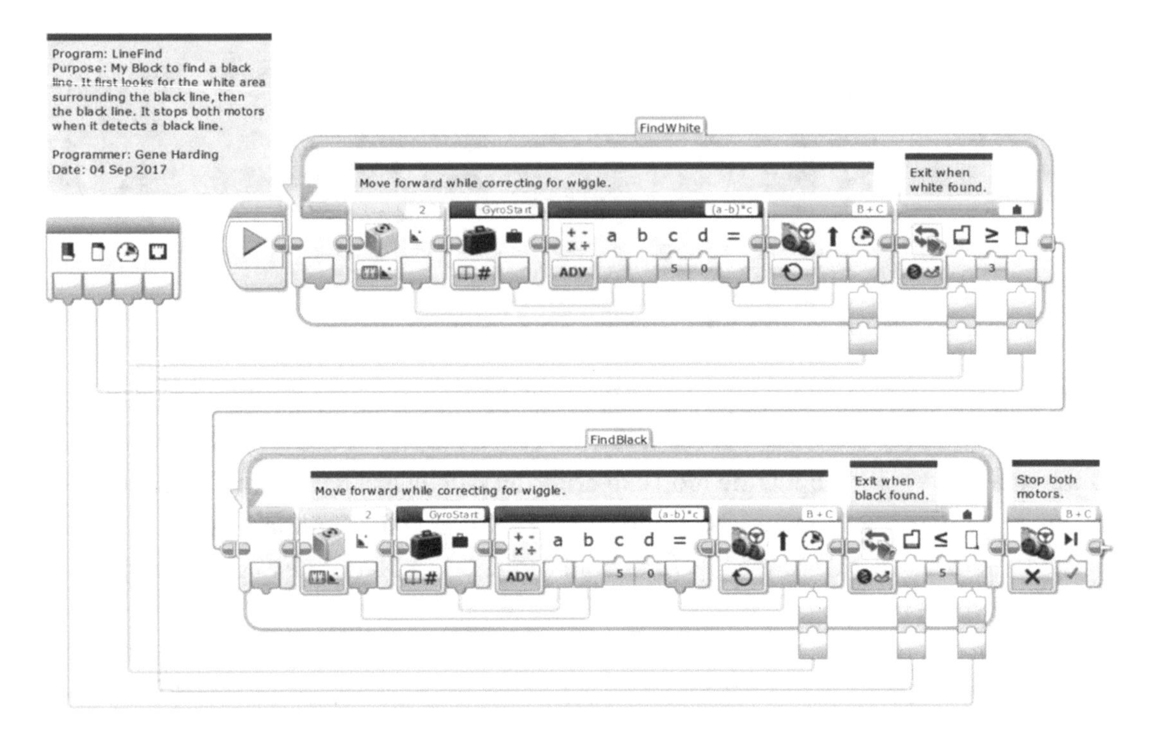

Figure 5-6. *My Block to find a black line*

Following a Black Line

Learning topics covered: My Blocks, unit conversion, feedback, looping constructs, proportional control

A color sensor can provide feedback to a Move Steering block so that the robot will follow the edge of a black line. Refer to Figure 5-7 as we explore this concept. First, note that the reflected light intensity over the center of the black line is at or close to 0, over the center of the white line it is about 100, and at the junction of the two it is about 50. Second, remember that the Steering input to a Move Steering block causes the robot to turn right when it is positive, turn left when negative, and go straight when zero. Thus, if the robot is following the edge of the black line and going straight, the +50 Reflected Light Intensity value from the color sensor must be translated to a 0 value for the Move Steering block. A Math block can do this operation quite easily by subtracting 50 from the Reflected Light Intensity value.

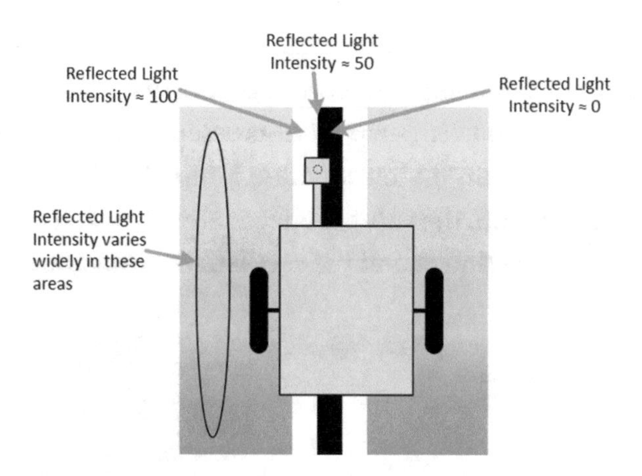

Figure 5-7. *Line following concept*

Now, think about what happens if the robot drifts to the left: The sensor is over more of the white line and less of the black line, which means the Reflected Light Intensity value increases to somewhere in the range of 51 to 100. After subtracting 50 the value is still positive, and a positive Steering value causes the robot to steer to the right, back toward the edge of the black line. Likewise, if the robot drifts to the right, the Reflected Light Intensity value decreases, and after subtracting 50 the value will be negative, which will cause the robot to steer left, back toward the edge of the black line.

Remember the proportional control technique described in Chapter 3 for wiggle control? This is the same control technique because the farther the robot drifts from the edge of the black line, the higher the steering correction to move it back. In this case, however, there is one more thing to consider: sensor placement. It might make sense to think of the robot as following the color sensor, so the sensor must be ahead of the driving wheels by at least a couple of inches, and preferably 3" to 5" inches or so. If the color sensor is placed too close to the driving wheels, it will be difficult or impossible to get the robot to follow the line smoothly.

Now it is time to start writing the line following program, which will follow a line for a specified distance. The first part of the program will be the same as the Forward2 My Block: Reset rotation sensors B and C, and convert distance from inches to degrees. Next will be a loop with the same exit criterion. Inside the loop will be a Color Sensor block in "Measure – Reflected Light Intensity"mode, followed by a Math block that subtracts 50 before passing that value to the Steering parameter input of a Move Steering block in On mode, as shown in Figure 5-8. Try the program and see how it works. Experiment with setting the light sensor slightly to the left and right of the edge of the black line.

What happens? The robot will probably oscillate several times before settling and following the line smoothly, or it might not settle at all. Try increasing the power from 20 to 30. At some point as the power (speed) is increased, the oscillations will not settle out. The problem is that the "gain" in the equation is too high. Because there is no gain parameter in the math equation, the value is effectively gain = 1. If a smaller value is used, the correction would be slower and the oscillations would be gentler.

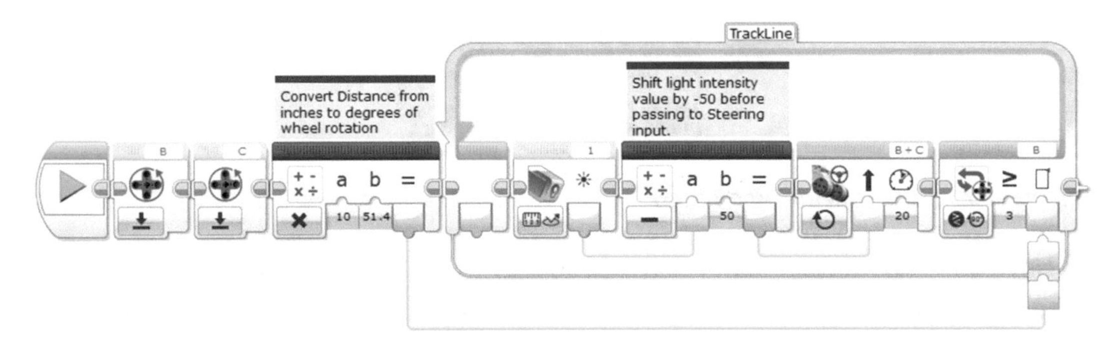

Figure 5-8. *Initial line following program*

Change the Math block mode to Advanced. Inputs "a" and "b" will be as before, but "c" will be used for the gain and "d" will be used to set which side of the black line to follow. The Steering parameter calculation will be

$$Steering = (a - b) * c * (-d) \tag{9}$$

Where

"a" is the reflected light intensity (data wire from Color Sensor block),
"b" is 50 (the amount to shift),
"c" is the gain, and
"d" is LineSide (1 for right side of black line, -1 for left side).

To start, set c = **0.5** and d = **1.0**, change the power back to **20**, then run the program again. Change d to **–1.0** and verify that the robot follows the left edge of the black line. Spend some time experimenting with values for gain (c input) and the Power input to the Move Steering block to get a feel for how they affect the line following performance. For a given speed, higher values of gain will allow the robot to track the edge of the line even if it is not initially aligned with it, but it will also tend to oscillate. Thus, there is a trade-off between smoothness and ability to track the line when approaching from an angle.

It is now time to make the My Block, with numeric inputs for LineSide, Port, and Distance. (The Port input will be useful after we add the second color sensor in the next section.) Although Gain and Power could also be set as inputs, it is probably a better strategy to make one version of the LineFollow program for going smoothly once the robot is aligned, and another version for dealing with poor initial alignment. That way the Gain and Power parameters can be tuned to optimize performance in each of those situations, and then left alone. The LineFollow My Block should look something like Figure 5-9.

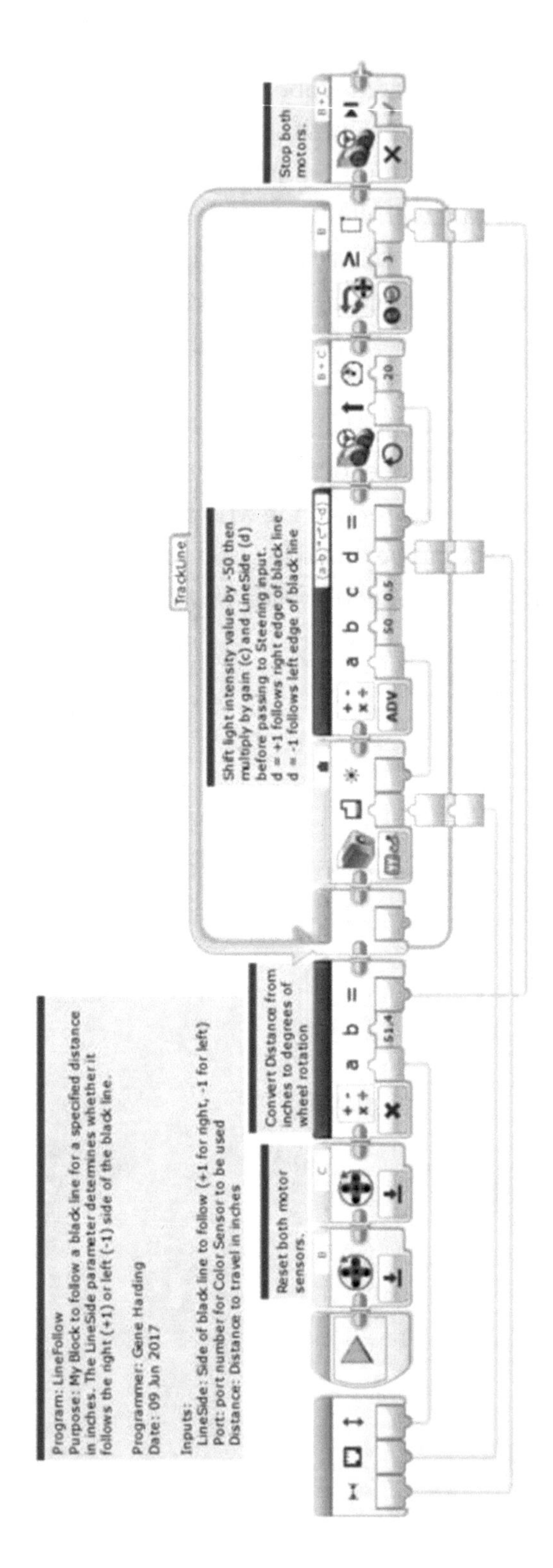

Figure 5-9. *LineFollow My Block*

Finding Black Line Intersections (Ts and Ls)

Learning topics covered: My Blocks, feedback, looping constructs

The concept of finding a line intersection is illustrated in Figure 5-10, which depicts the left color sensor being used to follow the left side of the black line as the right color sensor looks for the line intersecting from the right. Similarly, if looking for an intersecting line coming from the left, the right sensor would do the line following and the left sensor the detecting.

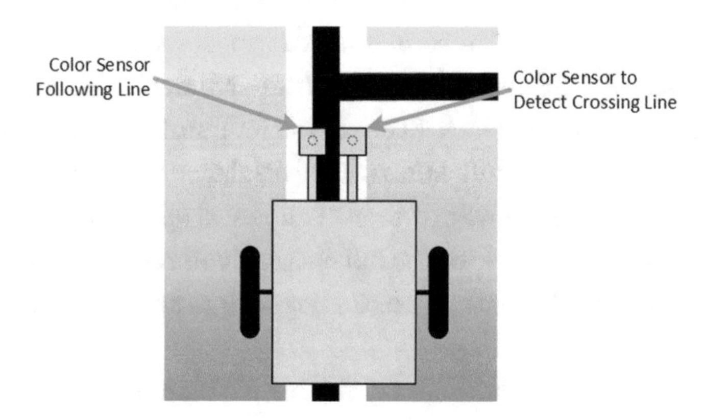

Figure 5-10. *Using two color sensors to find line intersections*

Performance will be more consistent if the color sensors are matched. Run the Calibrate2 My Block with the Port set to 1 in all four of the Color Sensor blocks, making a note of the minimum and maximum calibration values. Repeat with the Port set to 3 (or whichever port your other color sensor is using). If both values are within a few percent of each other, it should be fine. If not, it might be worth some effort to track down a color sensor that is a closer match to one of them.

The LineFollow My Block from the previous section is the basis for the LineTFind My Block. The inputs will be a little different, so a new My Block must be created from scratch. Start by copying LineFollow to a new program, then deleting the three blocks before the loop. These blocks will not be needed because the robot will keep going until it finds the intersecting line, regardless of the distance. The loop exit condition must also be modified to look for a light condition. Change its mode from "Motor Rotation - Compare - Degrees" to "Color Sensor - Compare - Reflected Light Intensity", and set the condition to **5** (≤).

There are now two sensor ports being used, but the side of the line to follow will determine which port is used to follow the line and which is used to find the intersecting line, so no port input(s) are needed. Create the LineTFind My Block with one numeric input called LineSide. Insert a Switch block before the loop, set its mode to "Numeric," and connect the LineSide input to it. On my robot, the left color sensor uses port 1 and the right sensor uses port 3, so if the input is 1 the robot will use port 3 to follow the right side of the line and port 1 to find the intersecting line. If the input is -1 the port assignments must be reversed.

We need two new variables to hold the port values, so create those variables, name them "FollowPrt" and "FindPort," or something that makes sense to you, and place one of each in the top part of the Switch block (in "Write - Numeric" mode), and one of each in the bottom part of the Switch block. (The reason for naming the first one "FollowPrt" instead of "FollowPort" is so the full name will be displayed when the variable appears in the program. If it is named "FollowPort" then "FollowP..." appears as the variable name because there is not enough space for the full name. If you cannot see the full variable name, just abbreviate it until you can. You can delete unused variables in the Project Properties window.)

The top value of the Switch block can stay at 1, but the bottom part must be changed from 0 to -1. Note that the "dot" selection next to the 1 at the top indicates the default, so if the input is any value other than -1 the Switch block will execute the top portion. When the input is 1, FollowPrt should be 3 and FindPort should be 1, and vice versa when the input is -1.

Next, place a copy of FollowPrt, in "Read - Numeric" mode, just before the Color Sensor block. Change the block's Port input from constant to wired, and connect FollowPrt to that input. Similarly, place a copy of FindPort just before the end of the loop and wire it to the Loop block's Port input. The LineTFind My Block is shown in Figure 5-11. Don't forget to add comments to it and test it to make sure it works properly.

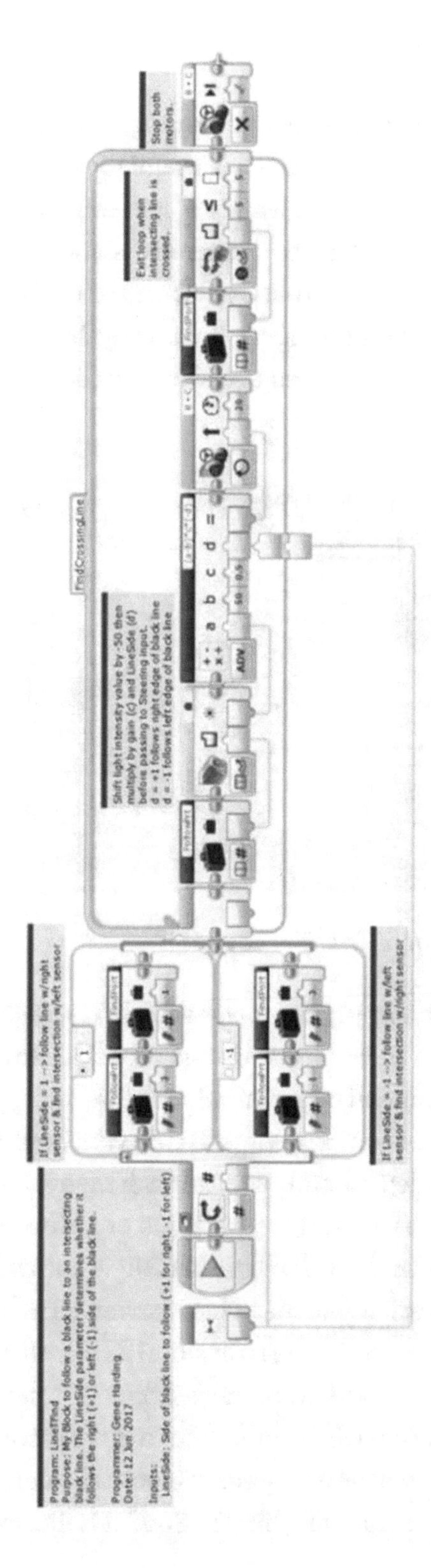

Figure 5-11. *LineTFind My Block*

Squaring Up on Lines

Learning topics covered: My Blocks, feedback, looping constructs

There might occasionally be times when it is useful to "square up" the robot so that it is perpendicular to a line. This is possible with two light sensors. The assumption for this My Block is that each light sensor is located the same distance forward of the driving axle, as shown in Figure 5-10. The concept is that the left sensor controls the left wheel and the right sensor independently controls the right wheel. They do this in a manner such that each sensor ends up positioned over the edge of the black line, as shown in Figure 5-12.

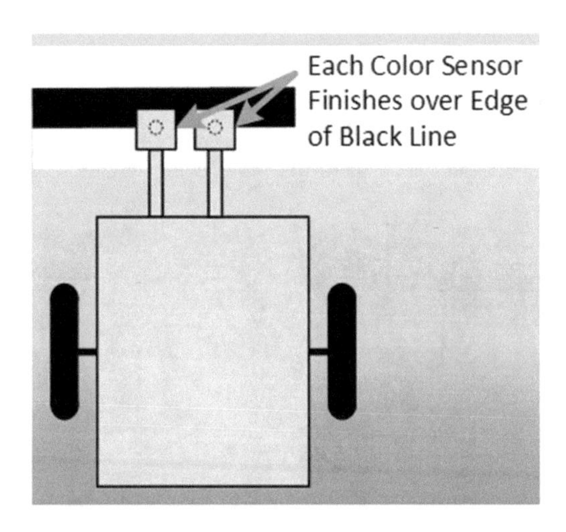

Figure 5-12. *Squaring up on a line*

Remember how the LineFollow My Block was set up? The Reflected Light Intensity value from the color sensor was shifted from a range of 0 to 100 to a range of –50 to +50 by subtracting 50, then it was multiplied by a gain value before being passed to the Steering input of a Move Steering block. A similar approach will work here, but a pair of Large Motor blocks will be used to independently control each motor, and the numeric values will go to the Power inputs (in lieu of the Steering input of a Move Steering block). Thus, when a sensor is over black the motor will move that wheel backward, and when the sensor is over white the motor will rotate that wheel forward. Two loops will execute in parallel, one for the left side and one for the right side. Ideally, after a couple of seconds or so each motor will settle into place with its corresponding light sensor over the edge of the black line. Each loop's exit criterion will be time, two to three seconds or so. Because this My Block will have no inputs, it will be a very small block. The name must therefore be very short to be visible, so let's call it something like "LnSqu." It is illustrated in Figure 5-13.

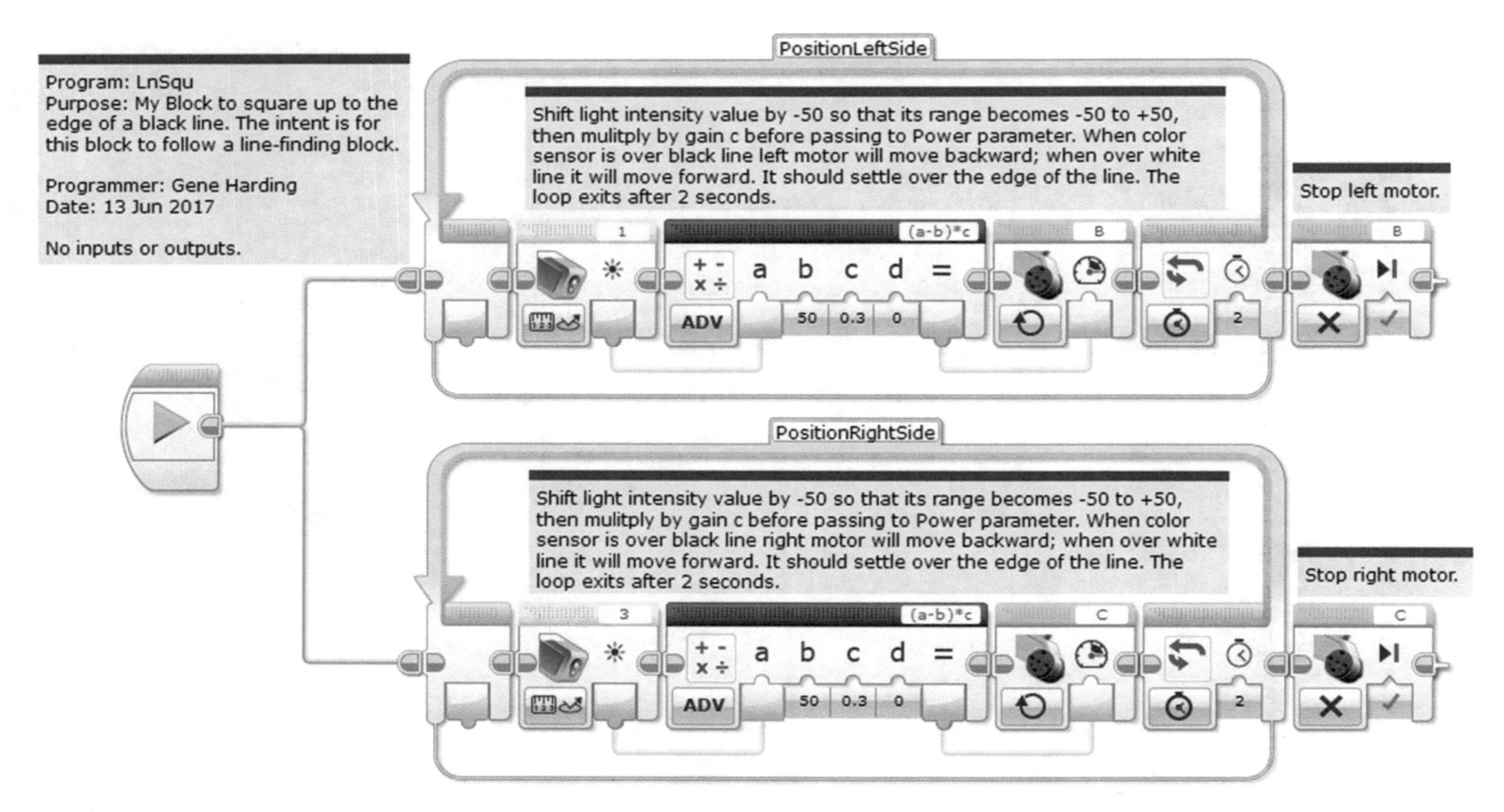

***Figure 5-13.** LnSqu (line-squaring) My Block*

Conclusion

You should now have a suite of several My Blocks that use lines to improve navigation accuracy. Using lines can make a very substantial difference, often the difference between success and failure on a given mission or set of missions. Walls can be used in a similar fashion to aid navigation, and that is the subject of the next chapter.

CHAPTER 6

Using Walls

LEARNING TOPICS

My Blocks, geometry, unit conversion, feedback, looping constructs, proportional control, and traction

REQUIREMENTS

1. Square up on walls.
2. Follow a wall for a specified distance while staying a constant distance from the wall without touching it.
3. Follow a wall for a specified distance while "leaning" against it.

Squaring Up on a Wall

Learning topic covered: geometry

Like lines, walls can be used to "square up" the robot. The process, however, is somewhat different. Probably the easiest way to square up on one of the border walls is simply to ram it with the front or back of the robot. As long as there are sturdy rectangular frames at the front and back of the robot, one of the forward or backup My Blocks can be used, using a Distance parameter set for slightly farther than the distance to the wall, and with a Power high enough to lose traction and generate wheel slip when the robot is blocked from moving. This action will cause the robot to

G. Harding, *Programming LEGO® EV3 My Blocks*, https://doi.org/10.1007/978-1-4842-3438-9_6

square up so that it is perpendicular to the wall. Because it is useful to have a "light guard" around the color sensors, the guard can serve the dual purpose of light shield and bumper. Figure 6-1 illustrates the concept, and Figure 6-2 and Figure 6-3 show our competition robot's front and rear bumper structures.

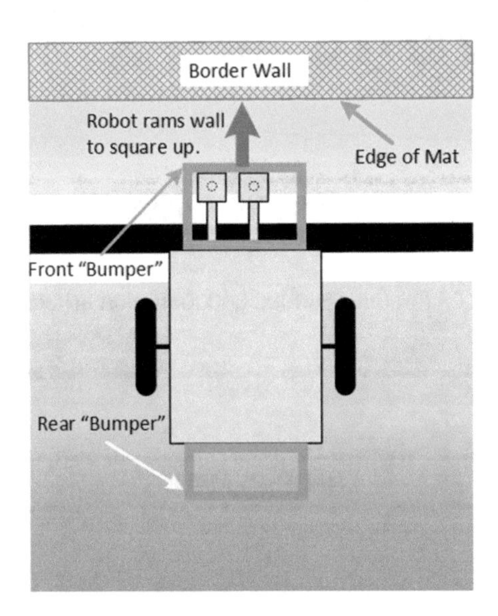

Figure 6-1. *Squaring up on walls*

Figure 6-2. *Robot front bumper*

Figure 6-3. *Robot rear bumper*

Following a Wall Without Touching It

Learning topics covered: My Blocks, feedback, looping constructs, proportional control

Sometimes the robot needs to navigate areas on a challenge board that do not have a black line, but are near a border wall that can be used as a navigation reference. This section addresses the case where the robot needs to follow along a wall while staying some distance away from it instead of touching it. The ultrasonic sensor can be used for this type of navigation.

The ultrasonic sensor uses sound waves at frequencies above the range of human hearing. In Presence mode it can listen for emanations from another sensor, but we use it in Measure mode. In Measure mode it emits ultrasonic waves and measures the time each wave takes to travel to the border wall and reflect back to the sensor. Because the speed of sound is known, if the time is measured precisely, the distance to the wall can be calculated. This process is called Sound Navigation and Ranging (SONAR), and is illustrated in Figure 6-4. Submarines have used SONAR for decades, and some marine mammals also have natural SONAR capabilities.

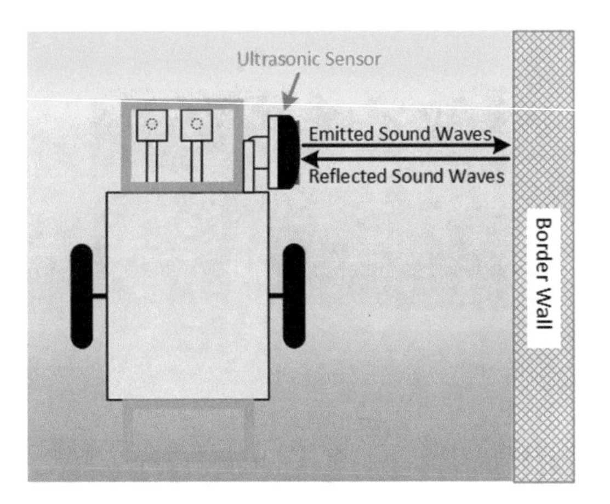

Figure 6-4. *SONAR concept*

The distance provided by the sensor can be used to compute a steering correction for a Move Steering block, similar to what we did with the LineFollow My Block using a color sensor. Note that, like the color sensor(s), the ultrasonic sensor must be mounted ahead of the drive axle for forward movement (behind the axle for backward movement). My robot uses port 4 for the ultrasonic sensor. Once your robot is configured, let's start programming!

An easy way to start is to copy the blocks from LineFollow into a new program. Replace the Color Sensor block with an Ultrasonic Sensor block in mode "Compare – Distance Inches," and set the correct port. Select all but the Start block and create a My Block named WallFollow with numeric inputs for distance to the wall and forward distance to travel (e.g., WallDist and FwdDist). Use the approach outlined near the end of Chapter 3 to tune the gain. For my robot I started with a gain of 1, increased it to 10, and then increased it to 100 before it began oscillating. After some experimenting a gain of 25 seemed to work best with my robot. An example My Block is shown in Figure 6-5.

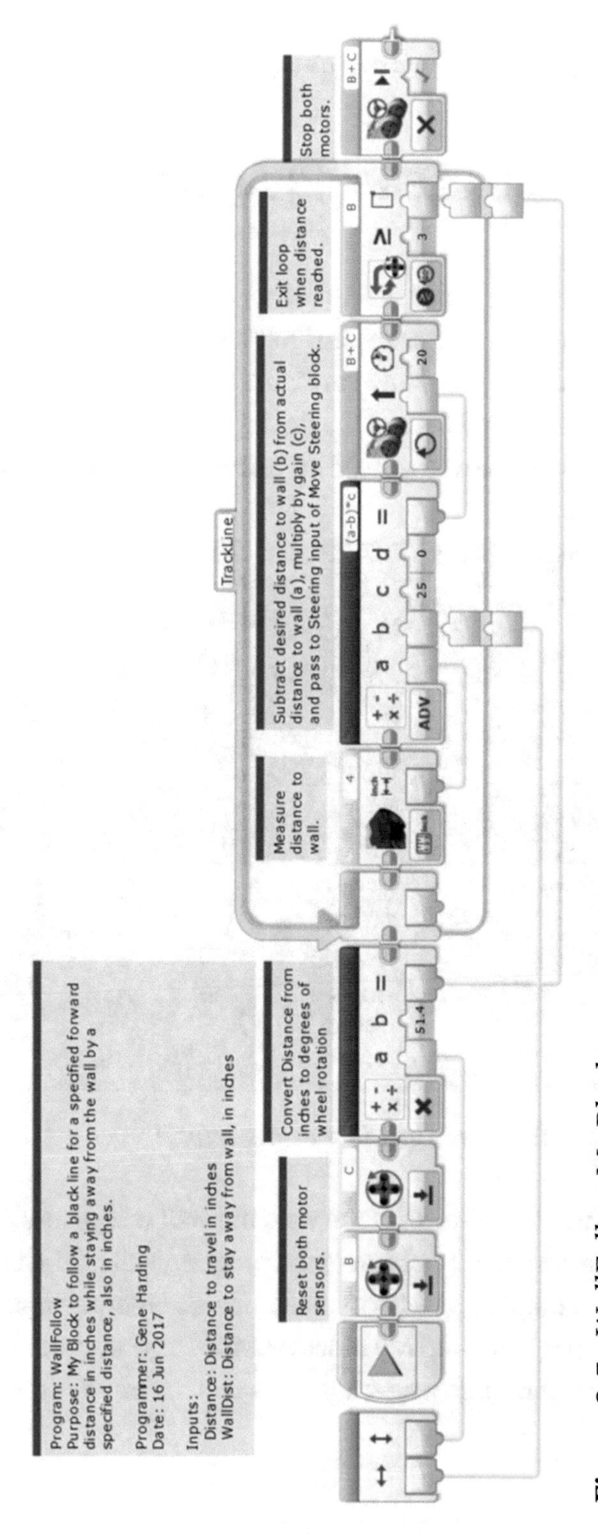

***Figure 6-5.** WallFollow My Block*

One final note: The ultrasonic sensor does not appear to work closer than 1.5" to 2.0" from a wall, likely due to the horizontal separation of the transmitter and receiver.

Following a Wall by Leaning Against It

Learning topic covered: geometry

Another possibility for wall following is to use wheels mounted to the side of the robot so that it can remain in contact with and "roll" along the wall. An example of a robot that uses wheels in such a fashion is shown in Figure 6-6. Note that if the wheels are not mounted far enough ahead of and behind the driving axle, the robot might rotate around the roller guide and turn into the wall instead of "leaning" against and following it.

Figure 6-6. *Robot with wheels to guide it along a wall*

Because the robot will "lean" against the wall, the wall will ensure the robot goes straight, and the wiggle correction portion of the program will not be needed. The leaning effect can be accomplished with a small nonzero Steering input to a Move Steering block. Our original Forward My Block provides a good starting point, but needs an additional input for steering, so a new My Block will have to be constructed using the My Block Builder.

Copy the two blocks from Forward into a new program file and make a new My Block with a name something like WallLean. Give it numeric inputs for Steering, Power, and Distance. The default value for Steering should be a low number. If the default will be a lean left, -2 is probably good, although a larger number might work if the roller is far enough ahead of the driving axle (or behind it if moving backward). As with the other forward My Blocks, 20 to 30 is probably a reasonable default value for Power, and perhaps 10 for the Distance parameter. Figure 6-7 shows an example of this My Block.

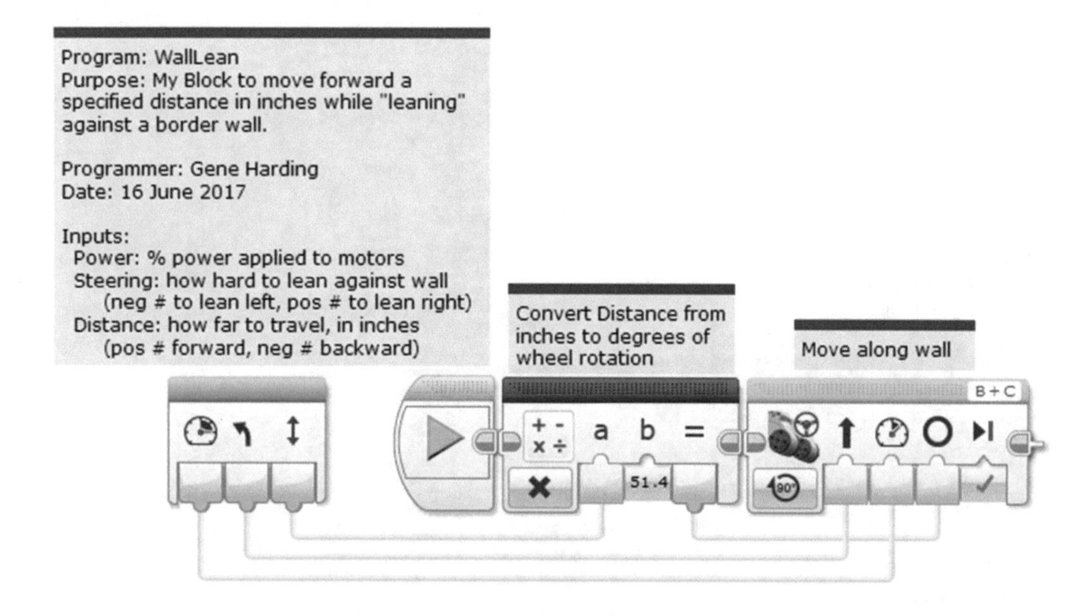

Figure 6-7. *WallLean My Block*

Conclusion

At this point, your robot should have a pretty full repertoire of My Blocks for navigating a competition board: moving forward and backward; turning; finding, following, and squaring up on lines; finding line intersections; and using walls. If you are willing and able to deal with the increased complexity, the next chapter deals with making it faster.

CHAPTER 7

Advanced Topics

LEARNING TOPICS

My Blocks, geometry, unit conversion, calibration, feedback, looping constructs, proportional control, variables, conditional constructs, torque, and traction

REQUIREMENTS

1. Move forward or backward a specified distance.
2. Ensure movement is straight (i.e., eliminate the "wiggle").
3. Enable optional "handoffs" so the robot does not have to stop at the end of the My Block.
4. Move as quickly as possible without wheel slip.

Handoffs: Continuing at Speed When the My Block Ends

Learning topics covered: My Blocks, conditional constructs

There are arguably two primary challenges to navigating a challenge board: precision and speed. The handoff technique deals with the constraint of time, and thus addresses the issue of navigation speed. Time is an important factor because each FLL Robot Game round only lasts 2.5 minutes, or 150 seconds. Saving a few seconds here and there can allow for more completed missions to achieve higher point totals.

G. Harding, *Programming LEGO® EV3 My Blocks*, https://doi.org/10.1007/978-1-4842-3438-9_7

To illustrate the usefulness of a handoff, consider the start of a mission set programmed by one of my former teams. The robot had to go forward more than 20", detect a line, curve to the left, detect another line, follow that line to its intersection with another line, then continue following the second line for a specified distance before making a right turn. It was necessary to stop before making the right turn, but not any time before. Imagine how the run would look if the robot stopped after every single My Block leading up to the turn: forward, stop, find line, stop, curve left, stop, find line, stop, follow line, stop, find line, stop, follow line, stop, turn right. All of these unnecessary stops waste time. If each stop wastes around 0.5 seconds, then just the beginning portion of the mission set would waste 3.0 seconds. Using the handoff technique saved several seconds per mission set, which allowed the team to add a few missions to their Robot Game strategy. The savings are even greater when acceleration and deceleration are used in the Forward and Backup My Blocks (covered later).

This technique is a feature to enable faster navigation that can be added to several My Blocks, such as Forward, LineFind, LineTFind, LineFollow, and WallFollow. For now, we start with Forward2 and make several changes to create a Forward3 My Block that can execute Handoffs at speed. The first change is to add something at the end of the program to allow the robot to either stop or continue at speed. This programming block is called a Switch block, shown in Figure 7-1. It has multiple "branches" where program flow can continue. The chosen path depends on the value of the input. In this case there will be two possible paths: one to tell the robot to stop and the other to tell the robot to continue at speed.

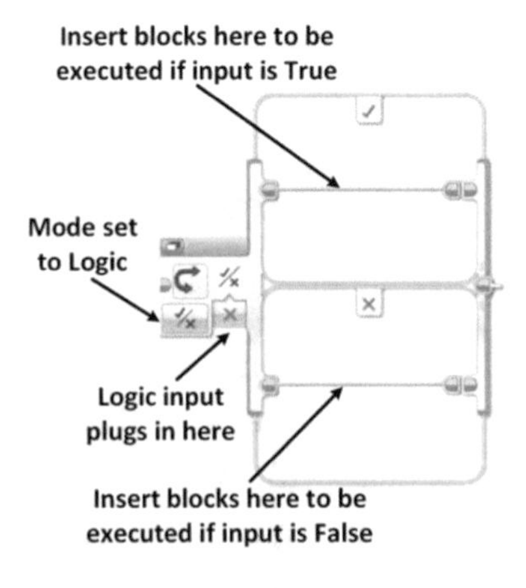

Figure 7-1. *Switch block*

The second addition will be an input to tell the robot which branch of the Switch block to follow. Because there are only two possibilities, a Logic input will work well (i.e., either True or False). Let's name the input BrakeAtEnd. If BrakeAtEnd is True, the robot will stop when it reaches the end of the My Block; if it is False, the robot will continue moving at its present speed and control will pass to the next program block.

The wiggle correction code creates an issue with Handoffs that will require more changes. Imagine a handoff between two My Blocks, both of which perform wiggle correction to keep the robot going straight. Now, suppose the first My Block is in the middle of doing a correction when it reaches the specified distance, so it is not pointed quite straight ahead. What happens when the next My Block begins? It loads GyroStart with the current direction and starts tracking that direction, which is slightly off. We can fix this problem by passing the GyroStart value from one My Block to the next.

Thus, the third change to make to Forward3 is to add a numeric output, GyroOut, to pass the GyroStart value to the next My Block. Likewise, a numeric input, GyroIn, is needed to accept the output of the previous My Block; and a logic input, Handoff, must be added to indicate that a handoff has happened. If RcvHandoff is True, GyroStart will be loaded with GyroIn from the previous My Block; if RcvHandoff is False, GyroStart will be loaded with the current direction just like in Forward2. Finally, the value of GyroStart must go to the output in case it is needed for a Handoff. The changes to create Forward3 from Forward2 are summarized here:

- Add logic input BrakeAtEnd to indicate whether robot should stop, or not, after executing this My Block.
- Add Switch block at end to either stop robot or continue at speed, depending on value of BrakeAtEnd.
- Add numeric output GyroOut to pass GyroStart value to next My Block.
- Add numeric input GyroIn to receive GyroStart value from previous My Block.
- Add logic input RcvHandoff to indicate whether the My Block is receiving a Handoff or not.
- Add a Switch block at the beginning to load GyroStart with either GyroIn or the current direction, depending on the value of Handoff.
- Add a GyroStart variable to the end of the program chain, in "Read - Numeric" mode, and wire its output to the GyroOut output of the My Block.

Because Forward3 will have new inputs and a new output, it cannot simply be copied from Forward2 in the Project Properties window. A new My Block must be created using the My Block Builder. The good news, however, is that the blocks can be copied from inside Forward2 so that the entire program does not have to be re-created from scratch. Begin by copying Forward2 in the Project Properties window, then go to the programming window for the new My Block, double-click the tab, and rename the new My Block "Temp." It will be deleted later.

In Temp, unplug the Distance input from the first Math block, and the Power input from the loop. Now the program blocks can be copied to a new program without leaving "hanging" connections from the My Block inputs. (It is safer to make a temporary program for this purpose than to modify the original My Block and redo the connections later.)

Next, create a new program, paste the blocks into the new program, and connect them to the Start block. Select everything but the Start block and select **Tools ➤ My Block Builder**. Name it Forward3 and select an appropriate icon. Add the following parameters:

- **BrakeAtEnd**: Parameter Type Input, Data Type Logic, Default Value True, and select an appropriate icon.
- **Distance**: Parameter Type Input, Data Type Number, Default Value 5, and select an appropriate icon.
- **Power**: Parameter Type Input, Data Type Number, Default Value 30, and select an appropriate icon.
- **GyroIn**: Parameter Type Input, Data Type Number, Default Value 0, and select an appropriate icon.
- **GyroOut**: Parameter Type Output, Data Type Number, and select an appropriate icon.
- **RcvHandoff**: Parameter Type Input, Data Type Logic, Default Value False, and select an appropriate icon.

Double-check to make sure everything is correct, then click **Finish** to create the My Block.

Now you are ready to set up Forward3 to allow Handoffs, but first drag the output block for GyroOut out of the way so it does not get covered up as you add programming blocks. Begin by adding a Switch block to the very end of the program, and set the mode to Logic. If BrakeAtEnd is true, the robot should stop, so drag the last Move Steering block (the one with mode set to Off) into the True branch of the Switch block, then use a data wire to connect the BrakeAtEnd My Block input to the Logic input of the Switch block. As before in Forward2, connect the Distance input to the "a" input of the first Math block and the Power input to the first Move Steering block (the one inside the loop).

A second Switch block is needed near the beginning of the program to determine what value to load into GyroStart. Place this Switch block before the first Math block, set its mode to Logic, and connect RcvHandoff to its control input. If RcvHandoff is False, the My Block needs to load GyroStart with the value from the gyro sensor, so drag the Gyro Sensor and Variable (GyroStart) blocks to the False branch of the Switch block. In the True branch of the Switch block, GyroStart needs to be loaded with GyroIn, but note that the Switch block will not allow you to connect a data wire when it is in "Flat View" (try it and see for yourself). It will allow this action when in "Tabbed View." The view can be toggled back and forth by clicking the upper left corner of the Switch block. Tabbed View only shows one of the multiple branches at a time, however, which I believe makes reading the program code harder. What we will do is always load the value of GyroIn into GyroStart, just before the Switch block, then either leave it alone if RcvHandoff is true or overwrite it with the current gyro sensor value if RcvHandoff is False. To accomplish this, just insert a Variable block for GyroStart, with mode set to "Write - Numeric," just before the Switch block, then connect GyroIn to it with a data wire. Finally, add a GyroStart Variable block to the end of the program chain, with mode set to "Read - Numeric," and wire its output to the GyroOut output of the My Block. The program should look something like Figure 7-2. Modify the comments appropriately and try it. Remember that there must be a block after Forward3 in the program to test whether the motors really continue running. A Wait block set for a time of a few seconds works well for this type of test.

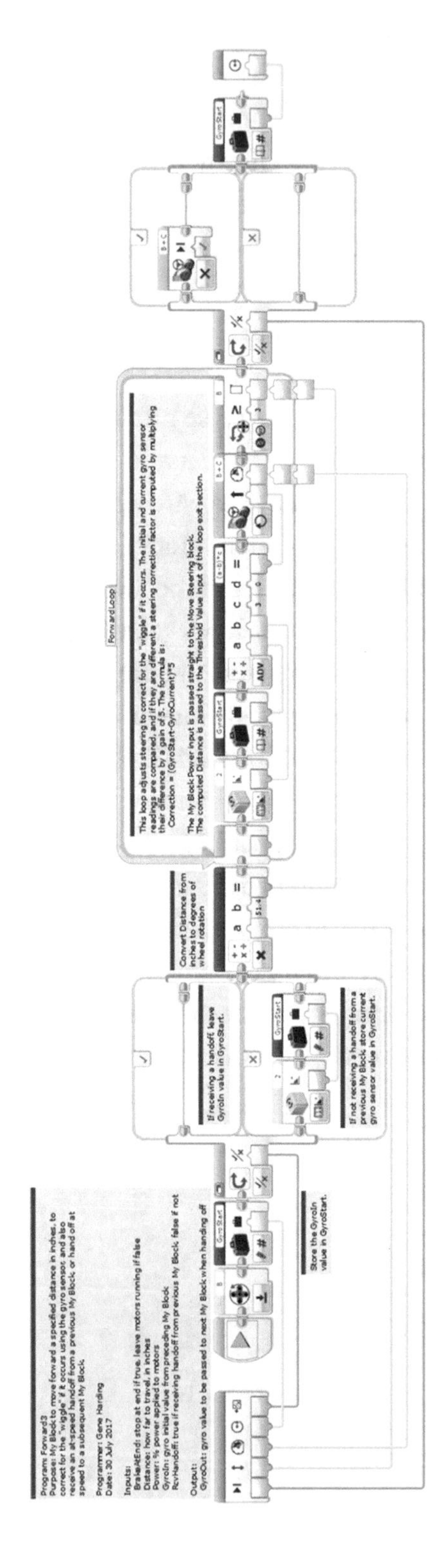

Figure 7-2. *Forward3 My Block with option to stop or continue running at end*

Forward3 can be copied, named Backup3, and modified with the same four changes that were made to convert Forward2 to Backup2:

- Negate the Distance conversion factor because the motor counters will be starting at zero and decrementing to negative values.
- Change the Loop block exit condition from "≥" to "≤".
- Reverse the steering correction by changing the formula from "(a-b)*c" to "(b-a)*c", or by negating the gain value, "c".
- Add a Math block before the loop to negate the Power input before it is passed to the Move Steering block inside the loop.

Of course, remember to adjust the comments appropriately.

Accelerating and Decelerating Using Forward3 My Blocks

Learning topics covered: My Blocks, torque, traction

This section and the next address the fourth requirement stated at the beginning of this chapter: Move as quickly as possible without wheel slip. The problem is traction. If the motor power is set to a starting value that is too high, the wheels or tires can slip (i.e., "peel out") as the robot picks up speed from a standstill. Likewise, the wheels can skid if the robot stops suddenly. The solution presented here is to accelerate smoothly to some maximum speed, then decelerate smoothly.

A relatively simple solution uses the Forward3 My Block from the previous section. Acceleration and deceleration will be implemented in discrete steps by cascading multiple Forward3 blocks together. The following example describes how to begin from a standstill, accelerate to a top speed, and then decelerate and stop. It uses several Forward3 blocks, and requires some trial and error to verify proper operation.

Begin with a Forward3 block with a starting power that will not generate wheel slip. For the robots that my team has used over the years, a power of 20 was usually safe, although this is a determination you will have to make experimentally. The following assumptions apply to this example:

- A starting power of 20 percent will not generate wheel slip.
- Power increases in steps of 20 percent will not generate wheel slip while the robot is moving.
- 1" is a sufficient distance for each acceleration step.
- Maximum desired power is 80 percent.
- Distance to move is 20".

The implementation involves seven Forward3 blocks. The first block uses a power of 20 for a distance of 1", with **BrakeAtEnd** set to False. The second and third blocks are the same, except that each uses a power 20 percent higher than its preceding block. The fifth, sixth, and seventh blocks each step power down 20 percent, and **BrakeAtEnd** is True for the final block. Because blocks 1 through 3 and 5 through 7 each specify a distance of 1" (6" total), the fourth block, running at 80 percent power, specifies the remaining 14" for a total distance of 20". This program is shown in Figure 7-3.

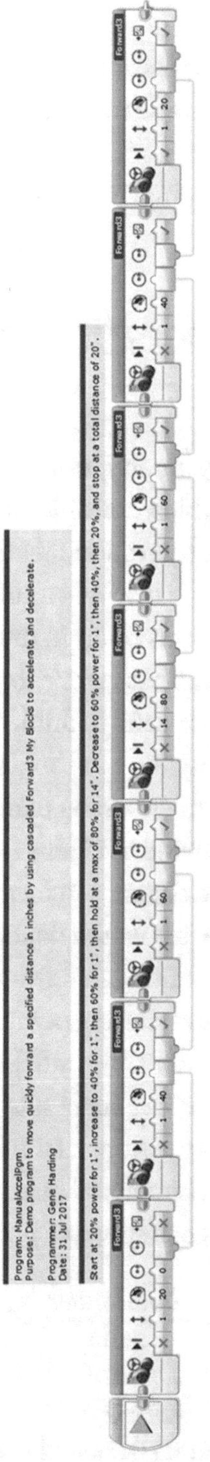

Figure 7-3. *Acceleration and deceleration implemented with cascaded Forward3 My Blocks*

If wheel slip occurs, which might only be noticeable by observing inconsistent distances, the acceleration and deceleration could be made less aggressive by using smaller power increments and decrements at each step. Alternatively, if quicker transit time is desired, the program could be made more aggressive by using larger power steps if it does not cause wheel slippage.

This approach to acceleration, although it works, is tedious to use inside a program and can result in very large programs.

Accelerating and Decelerating with a Single My Block

Learning topics covered: My Blocks, geometry, unit conversion, feedback, looping constructs, proportional control, variables, conditional constructs, torque, traction

Implementing acceleration and deceleration inside a single My Block results in cleaner programs, but the My Block itself involves a substantial increase in complexity over any of the programs developed to this point in the book. If done well, however, it should be simpler to use, smoother, and less error prone.

Before digging into the program, it is useful to think through what it must do at a conceptual level. Let us first consider the case of a robot beginning from a standstill, accelerating to some maximum speed, maintaining that speed for some distance, then decelerating to a stop. This is illustrated in Figure 7-4(a). Inputs to the My Block would include the maximum speed and distance to travel. The rate of acceleration and deceleration would be a constant value that would depend on factors like the wheel or tire size and texture of the mat surface. This value can be determined experimentally, but based on my experience with FLL robots, a good starting point is probably one wheel rotation, 360°, to accelerate from a stop to top speed. The motor blocks allow us to control power, which is essentially the same as speed, so let's substitute power for speed on the vertical axis of Figure 7-4(a). The acceleration rate is the slope of the line labeled Accelerate. The vertical distance is then 100 percent power. The horizontal distance is one wheel rotation, 360°. The slope of the line is thus the rate of change in speed or power, so we will call it ChgRate:

$$ChgRate = \frac{100\% \, Power}{360^\circ} = 0.278 \frac{\%}{^\circ} \tag{10}$$

We define ChgRate as a constant in the My Block, which we will name Forward4. For deceleration we negate ChgRate so that it is a negative number with the same magnitude.

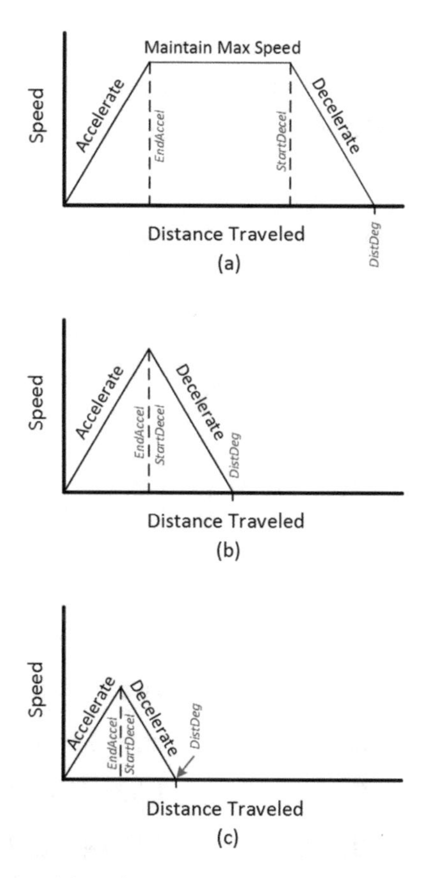

Figure 7-4. *Basic accelerate/decelerate plots*

There are three values in Figure 7-4(a) that must be calculated inside the My Block. They are depicted in red italics in the figure: EndAccel, StartDecel, and DistDeg. DistDeg is the same as Distance in the previous Forward My Blocks, except converted from inches to degrees. The value was calculated in previous My Blocks, so it only needs to be saved to a variable in the Forward4 My Block. EndAccel is the distance (in degrees) at which the robot reaches top speed (motor power), which is the same as the distance needed to accelerate. StartDecel is the point (in degrees) at which the robot must begin decelerating. Before finding this value, the distance needed to decelerate must first be calculated. We call this value DecelDist. Also, because the variable name StartDecel is too long to show completely inside the name window of a Variable block, the name is shortened to StrtDecel.

At-speed handoffs complicate the calculations a bit because the starting or ending powers might be nonzero. This introduces two more inputs: StartPwr and EndPwr.

The rest of the inputs and output for the Forward4 MyBlock will be the same as for the Forward3 My Block: MaxPwr (replaces Power), Distance, GyroIn, GyroStart, RcvHandoff, BrakeAtEnd, and GyroOut. The inputs, outputs, and key internal values are summarized in Table 7-1.

Table 7-1. *Forward4 My Block Parameters*

Parameter	Unit	Type
StartPwr	%	Input
MaxPwr	%	Input
EndPwr	%	Input
Distance	Inches	Input
GyroIn	Degrees	Input
RcvHandoff	Logic	Input
BrakeAtEnd	Logic	Input
EndAccel	Degrees	Internal, calculated
DecelDist	Degrees	Internal, calculated
StrtDecel	Degrees	Internal, calculated
DistDeg	Degrees	Internal, calculated
ChgRate	%/degree	Internal, constant
GyroStart	Degrees	Internal storage for GyroIn
GyroOut	Degrees	Output

Before we begin programming, it will be useful to think about what the Forward4 My Block must do at a high level. We do this by first listing the tasks to accomplish, then representing it visually with a flowchart. The tasks to be performed are as follows.

1. Input several values: StartPwr, MaxPwr, EndPwr, Distance, GyroIn (stored in GyroStart), RcvHandoff, and BrakeAtEnd.

2. Perform initial calculations for EndAccel, DecelDist, StrtDecel, and DistDeg.

3. Execute a loop to increase power (by ChgRate each loop) to some maximum value while adjusting steering to keep direction straight ahead.

4. Execute a loop to maintain maximum power, while adjusting steering, until time to start decelerating.

5. Execute a loop to decrease power to some minimum value while adjusting steering.

6. Either stop both motors or leave them running, depending on the value of BrakeAtEnd.

7. Store the value of GyroStart into GyroOut so it will be passed to the next My Block, in case it is needed.

A flowchart is a graphical representation of the procedures to be executed by a program. Figure 7-5 shows these steps illustrated in a flowchart. Notice the similarity of the sections in the gray shaded boxes. It looks like an opportunity to make a My Block that will be embedded inside this My Block! Is this possible? Yes, and in this case it makes a lot of sense.

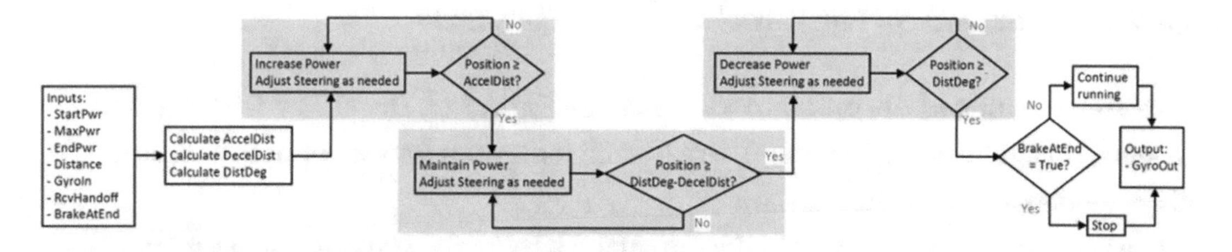

Figure 7-5. *Flowchart of Forward4 My Block*

Begin by creating a My Block with the appropriate inputs and output. The start of this My Block will look much like the Forward3 My Block, so select the first three blocks of Forward3 and open the My Block Builder window. Set up the following inputs and output:

- StartPwr, Input, Number, Default Value 0.
- MaxPwr, Input, Number, Default Value 60.
- EndPwr, Input, Number, Default Value 0.
- Distance, Input, Number, Default Value 12.
- GyroIn, Input, Number, Default Value 0.

- RcvHandoff, Input, Logic, Default Value False.
- BrakeAtEnd, Input, Logic, Default Value True.
- GyroOut, Output, Number.

The first part of Forward4 should look similar to Figure 7-6.

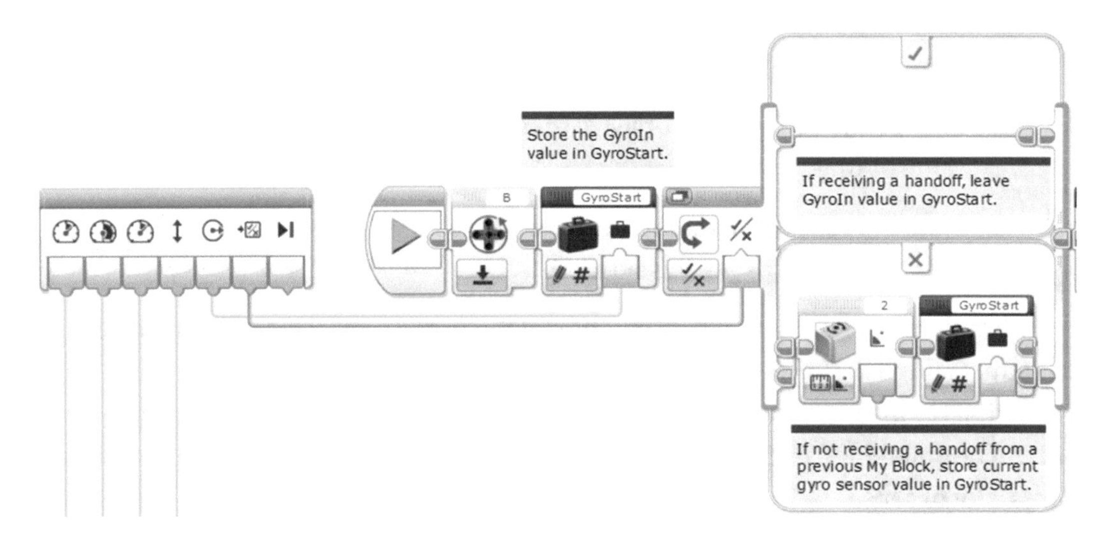

Figure 7-6. *First part of Forward4, similar to Forward3*

The next section of the program will calculate EndAccel, DecelDist, DistDeg, and StrtDecel, then determine whether there is sufficient room for the robot to accelerate and decelerate in the allotted distance.

The variables StartPwr and MaxPwr are used later in the program, so start by defining them. From the Data Operations programming tab, drag a Variable block to the end of the program chain. Name it StartPwr and set its mode to "Write - Numeric." Click the name box in its upper right corner and select Add Variable. Type **StartPwr** for the variable name and press Enter. Connect the My Block input StartPwr to its input with a data wire. Repeat the process to create a variable named MaxPwr, also wired to the corresponding My Block input.

Next, add the variable ChgRate. Set its mode to "Write - Numeric" and its value to **0.278**. Drag another Variable block to the end of the program chain, set its mode to "Read - Numeric," click the name box, and choose "ChgRate." Now we are ready to begin the calculations. Figure 7-7 illustrates the initialization of StartPwr and MaxPwr, as well as the calculations described later of EndAccel, DecelDist, DistDeg, and StrtDecel.

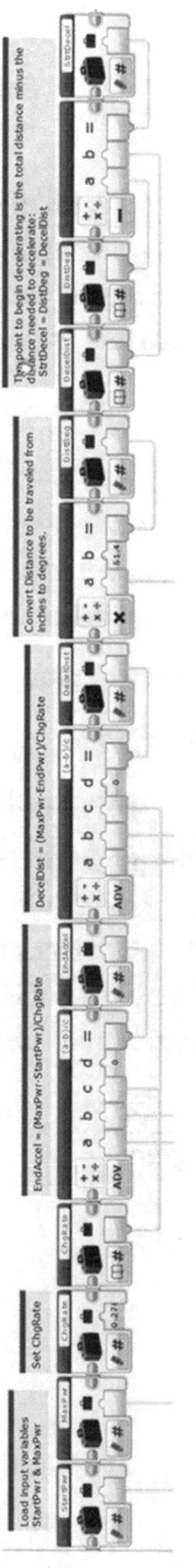

Figure 7-7. *Calculating EndAccel, DecelDist, DistDeg, and StrtDecel*

EndAccel and DecelDist can be calculated using the standard linear equation from basic algebra:

$$y = mx + b \tag{11}$$

where y is the final motor power (which correlates to robot speed), m is the rate of acceleration (+ChgRate) or deceleration (-ChgRate), x is the horizontal distance in degrees of wheel rotation, and b is the starting motor power (~robot speed).

Solving for x and substituting values to find EndAccel yields:

$$x = \frac{y-b}{m} \rightarrow EndAccel = \frac{MaxPwr - StartPwr}{ChgRate} \tag{12}$$

Equation 12 is straightforward to implement with an advanced Math block, so add a Math block to the end of the program chain, set its mode to "ADV," and set the equation to **(a-b)/c**. Use data wires to make the following connections to the Math block:

- "a" Math block parameter: MaxPwr input.
- "b" Math block parameter: StartPwr input.
- "c" Math block parameter: ChgRate variable.

Add a Variable block to the end of the program chain with mode set to "Write - Numeric." Click the variable naming box, select "Add Variable," and create a new variable named **EndAccel**. Wire the output of the Math block to the input of the EndAccel variable. This is illustrated in the first four blocks of Figure 7-7.

DecelDist can be found using the same approach as EndAccel:

$$DecelDist = \frac{EndPwr - MaxPwr}{-ChgRate} = \frac{MaxPwr - EndPwr}{ChgRate} \tag{13}$$

Add another Math block to the program, "ADV" mode, formula **(a-b)/c**, and use data wires to make the following connections:

- "a" Math block parameter: MaxPwr My Block input.
- "b" Math block parameter: EndPwr My Block input.
- "c" Math block parameter: ChgRate variable.

As with EndAccel earlier, add a Variable block to the end of the program chain, with mode set to "Write - Numeric." Click in the variable name box, select "Add Variable," and create a new variable named **DecelDist**. Wire the output of the Math block to the input of the DecelDist variable, as shown in Figure 7-7.

The distance to be traveled, in degrees, was calculated in Forward, Forward2, and Forward3. The difference in Forward4 is that the value will be stored into a variable named DistDeg. As in the previous My Blocks, add a Math block to the program chain with mode set to Multiply. Wire its inputs using data wires:

- "a" Math block parameter: Distance input.
- "b" Math block parameter: 51.4 (or whatever is appropriate to convert inches to degrees of wheel rotation for your robot).

Then, add a Variable block with mode set to "Write - Numeric," create a new variable named **DistDeg**, and wire the output of the Math block to its input, as shown in Figure 7-7.

Next, calculate the point where the robot must begin decelerating, StrtDecel. It is the difference between DistDeg and DecelDist:

$$StrtDecel = DistDeg - DecelDist \tag{14}$$

To do this calculation, add two Variable blocks to the program chain with mode set to "Read - Numeric": one for DecelDist and the other for DistDeg. Then add a Math block configured to Subtract, and another Variable block with mode set to "Write - Numeric" for parameter StrtDecel. Wire the first two variables to the Math block's input, and the Math block's output to StrtDecel, as shown in Figure 7-7.

The next task the program needs to perform is to consider the My Block's input parameters and determine whether they allow enough space to accelerate and decelerate. If there is, the program can continue normally, but what if there is not? How can we know, and how do we deal with that situation?

The test to determine if there is enough space is straightforward. Refer to Figure 7-4 again. In Figure 7-4 (a and b) there is sufficient room because DistDeg is at least as big as the sum of EndAccel and DecelDist, and that is the test:

$$DistDeg \geq AccelDist + DecelDist \tag{15}$$

If this condition is true, then the program can continue operating. Otherwise, as shown in Figure 7-4(c), there is not enough room and the program must somehow handle an error condition. In this case there are so many different situations that could lead to this error that it is probably wise to just flag it, stop the program, and let the user make appropriate adjustments to the My Block parameters.

To determine if there is enough room to accelerate and decelerate, we sum EndAccel and DecelDist, calculate DistDeg, compare those two numbers, and output a logic value to a Switch block. In the True branch of the Switch block, the program will continue normal operation. In the False branch it will flag an error and stop execution. Next, we add blocks to implement these steps.

Add a Variable block to the program chain, configured as "Read - Numeric," for the variable EndAccel. Do the same for the variable DecelDist. Then add a Math block in "Add" mode and connect the EndAccel and DecelDist variables to its inputs with data wires, as shown in the first three blocks of Figure 7-8.

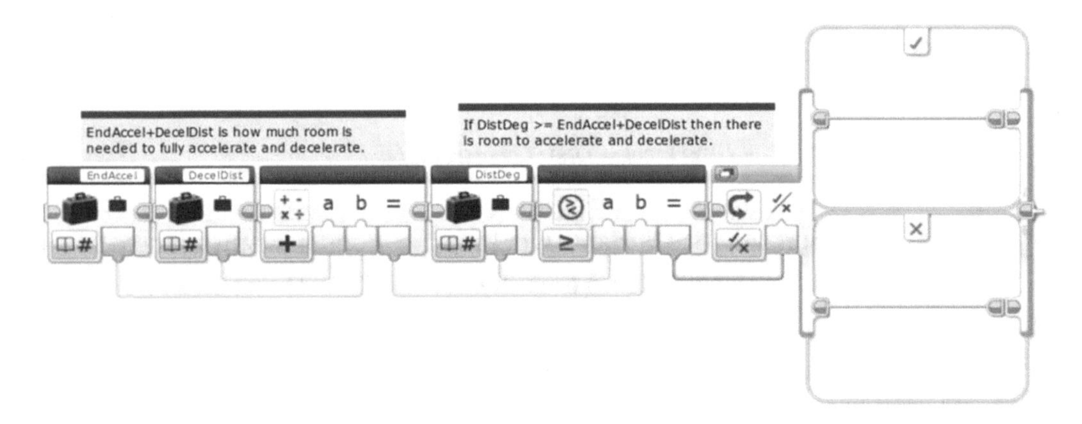

***Figure 7-8.** Determining if there is enough room to accelerate and decelerate*

Add a Variable block for DistDeg next, with mode set to "Read - Numeric," followed by a Compare block with mode set to "Greater Than or Equal To." This block will perform the test to determine if there is enough room to accelerate and decelerate. Connect the DistDeg Variable block output to the Compare block "a" parameter input with a data wire. Connect the "Add" Math block, the output of which is EndAccel+DecelDist, to the Compare block "b" parameter input. Finally, add a Switch block to the program chain, with mode set to Logic, and connect the Compare block output to the Switch block input with a data wire, as shown in Figure 7-8.

The True branch of the Switch block will contain instructions to make the robot accelerate, maintain speed as needed, and decelerate. The False branch of the Switch block will contain code to handle situations when there is not enough room to accelerate and decelerate fully. Let's deal with the False branch first. When there is not enough room to accelerate and decelerate, we need to provide clear feedback to the user and terminate the program early. The feedback will take two forms: The brick status light will turn red and an error message will be displayed. We will also turn off the drive motors before exiting the Switch block.

Turning the brick status light to red is easy. From the Action tab, drag a Brick Status Light block into the False branch of the Switch block, set its mode to "On," Color to "2" (red), and Pulse parameter to "False." Next, add a Move Steering block with mode set to "Off" and Brake at End parameter set to "False," as shown in Figure 7-9. (Turning off Brake at End will allow the robot to coast down instead of suddenly halting, which is probably better if it happens to be at a high motor power.)

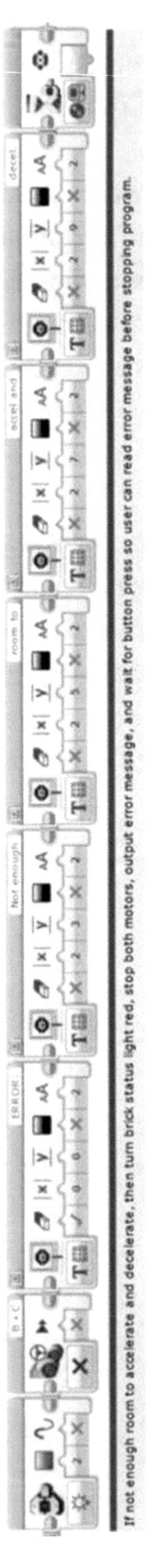

Figure 7-9. *Code to turn brick status light red, stop motors, and output error message*

A good error message should be clear, so we will use several lines to describe the problem. The message should be something like the one shown in Figure 7-10. Because this message is five lines it will take five Display blocks. Add a Display block right after the Move Steering block, set its mode to "Text - Grid," and set the display text (upper right box) to "ERROR:". Set its parameters as follows:

- Clear Screen: True.
- x: 0.
- y: 0.
- Color: False.
- Font: 2.

```
ERROR:
  Not enough
  room to
  accel and
  decel.
```

Figure 7-10. *Error message*

Add four more Display blocks to the program chain with all of their modes set to "Text - Grid," Clear Screen and Color set to "False," x set to "2" (to indent the error description), and Font set to "2." For the first one set the Text to "Not enough" and y to "3"; the second block's Text to "room to" and y to "5"; the third block's Text to "accel and" and y to "7"; and the last block's Text to "decel." and y to "9." The code in the False branch of the Switch block should look something like Figure 7-9.

As depicted in Figure 7-5, when the program is operating normally the next phases will be to accelerate to top speed, maintain that speed if needed, and decelerate to final speed. The same code can be used for all three functions, making it perfect for a My Block. This My Block will need three inputs: starting power, acceleration or deceleration rate, and distance at which to stop accelerating or decelerating.

Because the new My Block will use the wiggle correction code, we can take that code from Forward3. Copy all of the blocks from Forward3 (except the Start, Input, and Output blocks) to a temporary program and delete the blocks before and after the loop. There might be a short black line protruding from the bottom of the Move Steering block's power input, indicating that the software is confused about how to handle the missing connection. (The My Block input parameters did not come with the code when it was copied, so the connection to the power input was broken.) To remedy this, change the Move Steering block's mode from On to Off and back to On, then reconnect the data wire from the Math block to the Move Steering block's Steering input. At this point the new program should look like Figure 7-11. Now we need to add code to increase the power steadily.

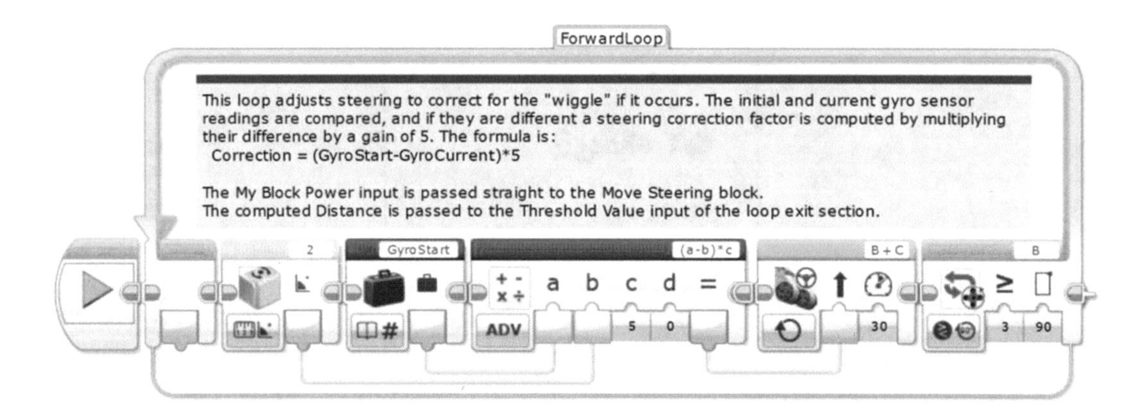

Figure 7-11. *Start of ChgFwdSpd My Block*

The new code will go inside the loop (ForwardLoop in Figure 7-11). Its function is to increase the motor power as the robot moves forward. Each time the loop executes, it will calculate a new power based on the distance the robot has traveled since entering the loop. Because the power needs to increase (or decrease, as the case may be) as the robot moves forward, it makes sense to mathematically link the power value to the robot's distance traveled, as shown in Equation 16.

$$NewPwr = StChgPwr + ChgRate * (CurrentDist - StDist) \quad (16)$$

where

NewPwr is the new power value for Move Steering block, in percent;
StChgPwr is the starting change power before the loop started executing, in percent;
ChgRate is the rate at which power will change, in percent per degree of wheel rotation;
CurrentDist is the total distance traveled, in degrees of wheel rotation; and
StDist is the distance traveled before the loop started executing, in degrees of wheel rotation.

Thus, the quantity (CurrentDist – StDist) is the distance the robot has traveled since beginning loop execution. Multiplying that quantity by ChgRate gives the amount of power to add to the starting power, StChgPwr, and it increases steadily as the robot moves forward. StChgPwr and ChgRate will be inputs to the My Block. StDist can be read from a motor rotation sensor before the loop begins, and CurrentDist can be read from a motor rotation sensor each time the loop executes. Because the NewPwr formula involves four parameters, it can be done with a single math block in Advanced mode. The first part of the loop should look something like Figure 7-12. Note that variables have been created for StDist, ChgRate, and StChgPwr. These variables must also be inserted in front of the loop, where they are in Write mode so they can be loaded with values. Inside the loop they are in Read mode so their values can be passed to the Math block. Using variables in this fashion keeps the data wires inside the loop rather than crossing the loop boundary, as shown in Figure 7-13.

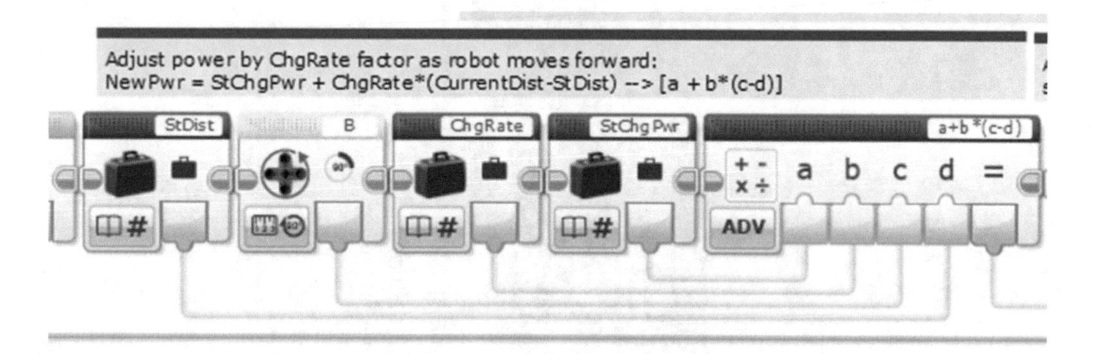

Figure 7-12. *First part of loop in ChgFwdSpd My Block*

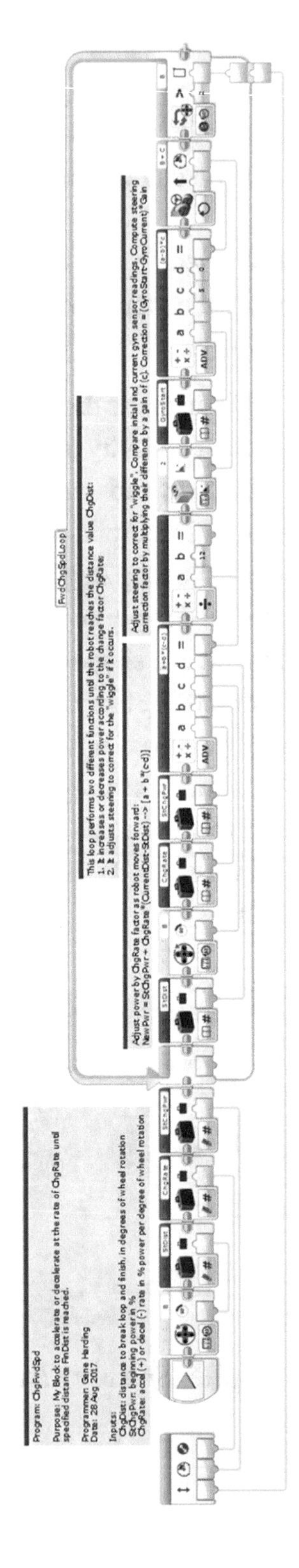

Figure 7-13. *ChgFwdSpd My Block*

You might want to manually type values into the data variables (and also the loop Threshold value) to experiment with the program before you convert it to a My Block. Once you are ready, create the My Block ChgFwdSpd with inputs ChgDist, StChgPwr, and ChgRate. It should look something like Figure 7-13. Take some time to experiment with using ChgFwdSpd in a program before moving on.

There is a new issue that might show up in the ChgFwdSpd My Block. Recall that we tuned the gain in Forward3 using a relatively low power (~30 percent) for the drive motors to prevent wheel slippage. If the robot tracks well and the steering is stable across a wide range of speeds, a fixed gain is fine. If the robot does not track well or it exhibits steering oscillations at some speeds, it might be appropriate to vary the gain with speed: higher gain for higher speeds, lower gain for lower speeds. A Math block, inserted after the Math block that computes drive motor power, can do this. Set the Math block to Divide mode, connect the computed motor power to the "a" input, and set the "b" input to a value that allows the robot to track without steering oscillations. For my robot a value of 15 worked well. A little experimentation should reveal a value that works for your robot.

Forward4 is almost done. The next thing to do is add the programming blocks to populate the True portion of the Switch block. Take another look at the flowchart of Figure 7-5. Three ChgFwdSpd My Blocks are needed: one to accelerate, one to maintain top speed, and one to decelerate. Start with the blocks to accelerate. These blocks will start the B and C motors at power of StartPwr, and increase the power at a rate of ChgRate until it reaches the distance EndAccel. In the True branch of the Switch block, add the three variables ChgRate, StartPwr, and EndAccel, all in mode "Read - Numeric." Then add a ChgFwdSpd My Block and wire the variables to the corresponding inputs of ChgFwdSpd, as shown in Figure 7-14.

Figure 7-14. *Blocks in Forward4 to accelerate*

The next three blocks will maintain top speed by holding the B and C motors at MaxPwr until it is time to begin decelerating at the distance of StrtDecel. Add MaxPwr and StrtDecel to the program chain in mode "Read - Numeric," then insert a ChgFwdSpd My Block. The parameter ChgRate should be 0, so set that value manually, then wire the two variables to the appropriate ChgFwdSpd inputs, as shown in Figure 7-15.

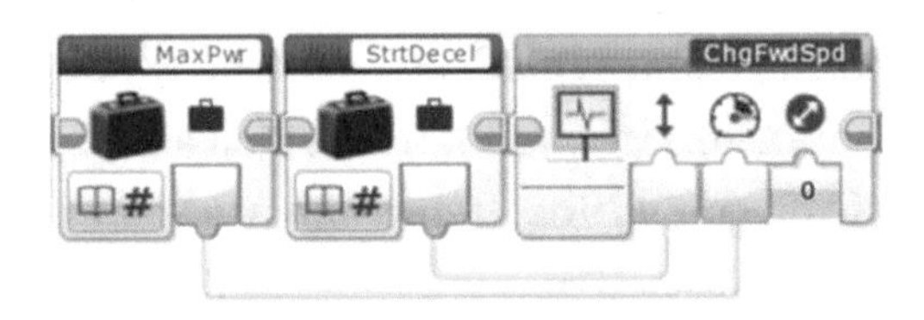

Figure 7-15. *Blocks in Forward4 to maintain top speed*

The last several blocks will decelerate the robot from MaxPwr at a rate of -ChgRate until DistDeg is reached. An extra Math block is needed to negate ChgRate, so add the variable ChgRate in mode "Read - Numeric" to the program chain, followed by a Math block in Multiply mode. Wire ChgRate to the Math block's "a" input, and set the "b" input to a value of **-1**. Insert MaxPwr and DistDeg variables and a ChgFwdSpd My Block. Wire the Math block output to the ChgRate input of the ChgFwdSpd My Block, DistDeg to the ChgDist input, and MaxPwr to the StChgPwr input, as shown in Figure 7-16.

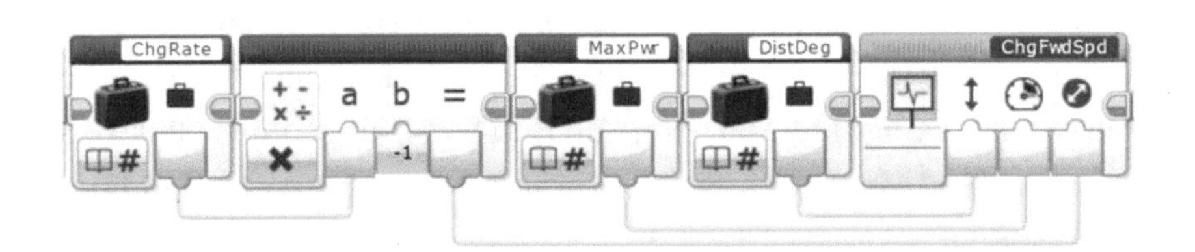

Figure 7-16. *Blocks in Forward4 to decelerate*

The entire Switch block with True program chain to accelerate and decelerate, and False program chain to turn the brick status light red, stop both motors, and output an error message is shown in Figure 7-17.

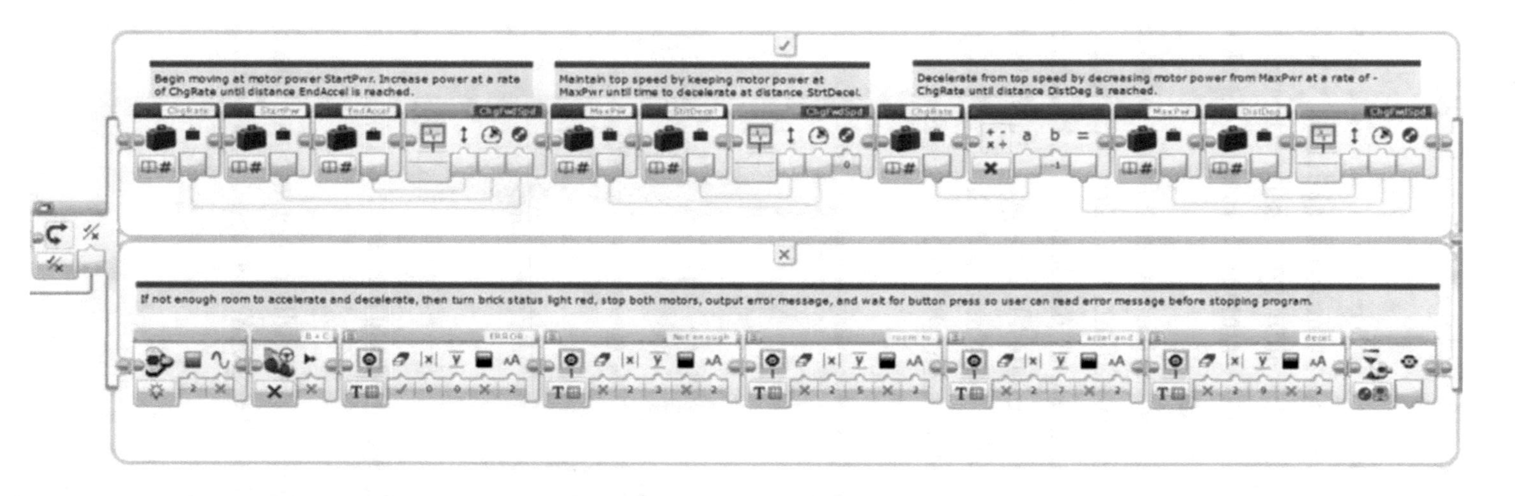

Figure 7-17. *Switch block to accelerate and decelerate (True branch) or stop robot and indicate error (False branch)*

Only two things remain: stopping both motors if BrakeAtEnd is True, and outputting the value of GyroStart (to the My Block output GyroOut) in case there is another My Block following Forward4.

First insert a variable block at the very beginning of the program with mode set to "Write -Logic," and set the variable name to **BrakeAtEnd**. Wire the My Block input BrakeAtEnd to this variable. Note that although they have the same name, the My Block input parameter BrakeAtEnd is not the same as the variable BrakeAtEnd, which is why the former must be wired to the latter.

Next, add another variable block to the very end of the program, with mode set to "Read - Logic" and name set to BrakeAtEnd. Insert a Switch block in Logic mode, and wire the BrakeAtEnd variable to its input. Inside the True branch of the Switch block, insert a Move Steering block set to mode Off. Finally, add a Variable block set to mode "Read - Numeric" and name set to GyroStart, and wire it to the GyroOut output of the My Block. The end of the Forward4 program should look like Figure 7-18.

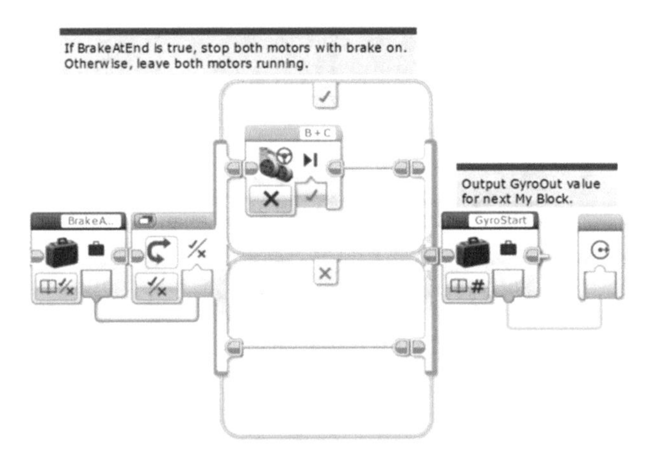

Figure 7-18. *Last part of Forward4 My Block*

Remember to add comments. They will be very important in a My Block as large and complicated as Forward4. The full Forward4 My Block is shown in Figure 7-19.

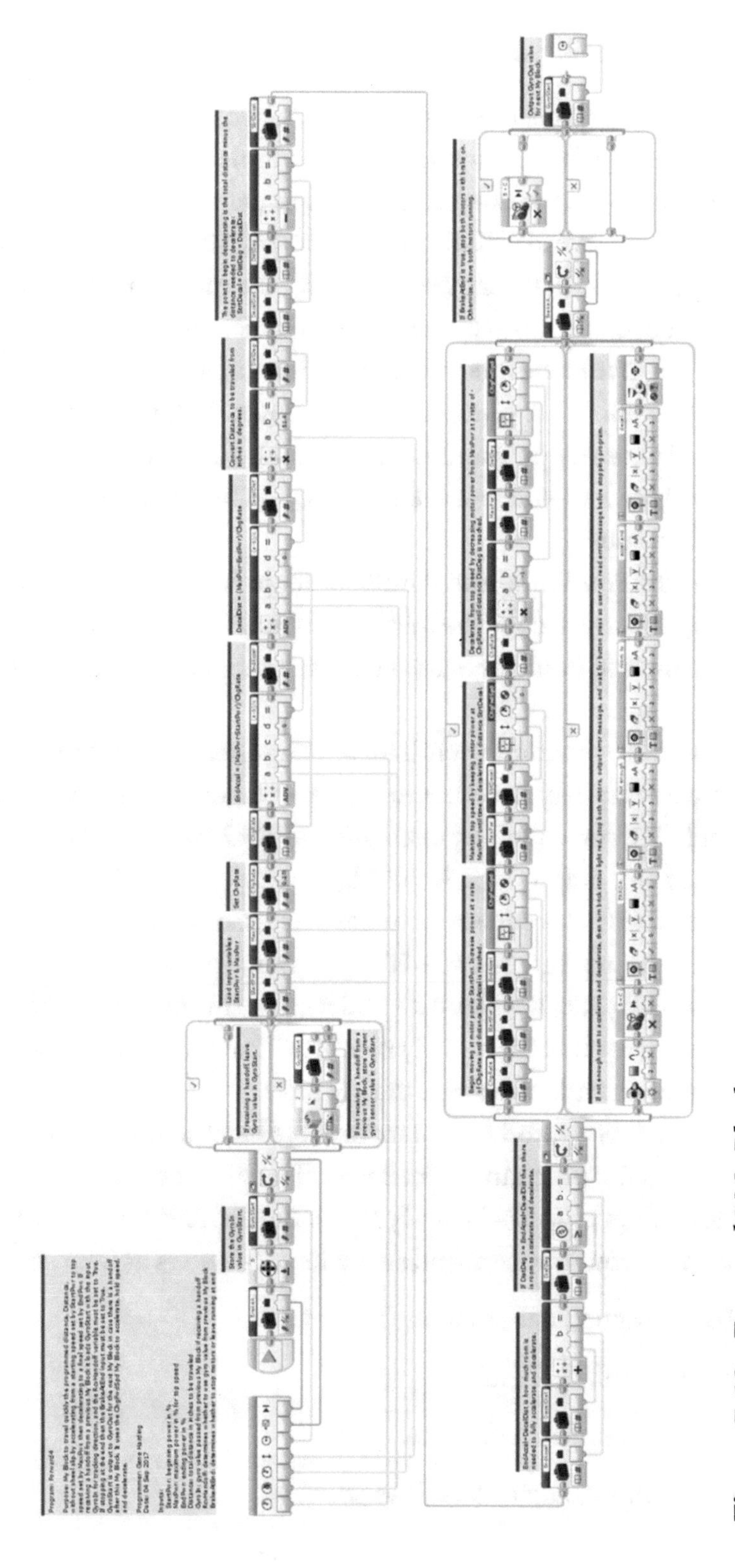

Figure 7-19. *Forward4 My Block*

Adding Handoffs to LineFind

Learning topics covered: My Blocks, feedback, looping constructs, conditional constructs

Handoffs at speed are probably most useful when transitioning from one type of My Block to another. For instance, the black lines on a competition mat can be very useful as is, but what if one wants to find a line and keep moving, instead of stopping? If the handoff feature implemented in the earlier My Blocks is implemented in the LineFind My Block, then the robot could move forward until it is close to the line (e.g., using a Forward4 My Block), then find the line to get a position fix (waypoint), and hand off to another Forward4 My Block without ever slowing down. This approach would not only eliminate wasted time stopping and starting, but also allow the robot to move at a relatively high speed, saving even more time. Let's modify the LineFind My Block to incorporate handoffs at speed.

Begin by copying the FindWhite and FindBlack loops from LineFind and pasting them into a new program. Next, copy the first three blocks (Motor Rotation, GyroStart Variable, and Switch blocks) from Forward3 and insert them just before the FindWhite loop in the new program.

Because the robot will sometimes continue moving after it finds the black line, it will be helpful to have some user feedback to indicate whether the robot detects the line when it passes over it. We can do this by changing the brick status light to a different color. From the Action tab, drag a Brick Status Light block to the end of the program chain, right after the FindBlack loop. Set its mode to On, Color to 1 (orange), and Pulse to False. As this block is almost at the end of the program, you will need to leave the light on or it will flash so quickly that it will not be noticeable. This means that the calling program (the program that contains this My Block) will have to turn it back to flashing green. We will return to that after we finish the My Block.

Now, copy the Switch block and GyroStart Variable block at the end of Forward3, and paste them to the end of the program chain. Select all of the blocks in the new program, except the Start block, then go to **Tools ➤ My Block Builder**. Name the My Block **LineFind2**, select an appropriate icon, and set up its parameters as follows:

- BrakeAtEnd: Input, Logic, default True.
- Port: Input, Number, default 1.
- Power: Input, Number, default 30.
- RcvHandoff: Input, Logic, default False.

- GyroIn: Input, Number, default 0.
- GyroOut: Output, Number.

Remember to pick an appropriate icon for each parameter, double-check everything, and click Finish to create LineFind2. Finally, wire the inputs and outputs as follows:

- GyroIn to the first GyroStart variable block.
- RcvHandoff to the input of the first Switch block.
- Power to the Power input of each Move Steering block.
- Port to the Port input of each Loop block.
- BrakeAtEnd to the input of the last Switch block.
- The output of the final GyroStart Variable block to the GyroOut output of the My Block.

Of course, don't forget to properly comment your new My Block, which should look something like Figure 7-20.

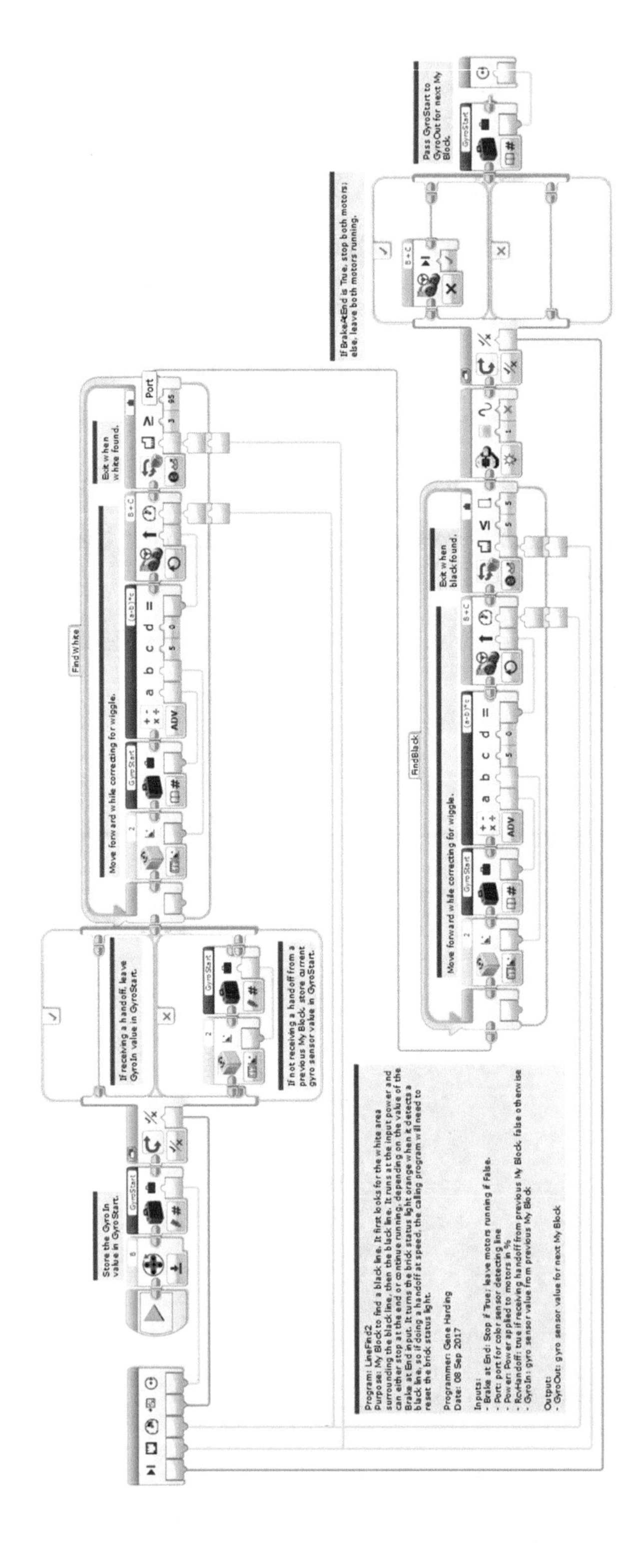

Figure 7-20. *LineFind2 My Block with handoff capability*

Now, let's write a brief test program to make sure the handoff function works properly, and to see how to handle the orange light. Recall that the brick status light normally flashes green while a program is running. A good practice for dealing with the light is to add code to the calling program that leaves the light orange for a second or two, then changes it back to flashing green. That way you will know that the program is operating normally until the status light is changed to orange or red by another program block.

Create a new program and insert a Forward4 My Block, then a LineFind2 My Block, and last a Forward4 My Block. Set the My Blocks' parameters as follows:

- First Forward4: StartPwr 20, MaxPwr 100, EndPwr 100, Distance 12, GyroIn 0, RcvHandoff False, BrakeAtEnd False, and wire GyroOut to the GyroIn input of the LineFind2 My Block.
- LineFind2: BrakeAtEnd False, Port 1 (or whatever port is used for the color sensor), Power 100, RcvHandoff True, and wire GyroOut to the GyroIn input of the last Forward4 My Block.
- Last Forward4: StartPwr 100, MaxPwr 100, EndPwr 20, Distance 8, RcvHandoff True, and BrakeAtEnd True.

Dealing with the status light is a perfect time to use parallel program execution. Instead of adding the following blocks to the end of the program chain, add them underneath the last Forward4 My Block, connected to each other but not the rest of the program:

- Wait block: Mode Time, Seconds 1 (or 2).
- Brick Status Light block: Mode On, Color 0 (green), Pulse True.
- Wait block: Mode Time, Seconds 2.

Grab the Sequence Plug Exit of the LineFind2 My Block and drag it to the Sequence Plug Entry of the first Wait block. The test program should look something like Figure 7-21. Position the robot more than a foot away from a black line and run the program. The robot should accelerate smoothly to top speed, navigate straight to the black line, turn the status light orange while continuing at top speed, decelerate and stop 8" after the black line, and wait for a couple of seconds with the green light flashing.

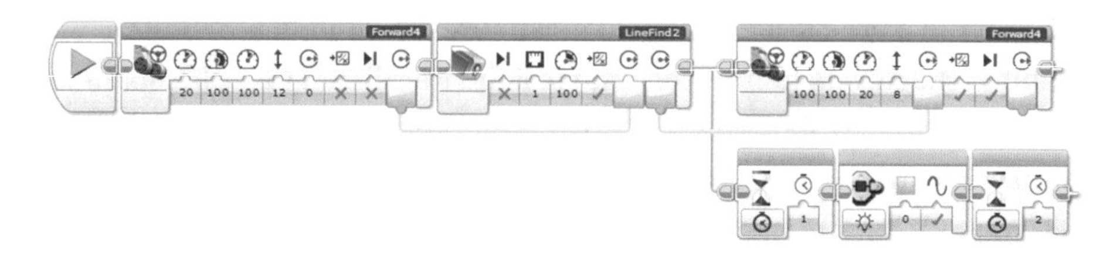

Figure 7-21. *Handoff test program*

The approach used in this section can be used to add Handoffs to other programs, such as LineFollow, LineTFind, WallFollow, and WallLean. Once you do this you will have a fairly complete and potent suite of My Blocks to enable navigation to pretty much any point on the game board, allowing you to focus more on the mission programming.

Combo Program

Learning topics covered: My Blocks, looping constructs, conditional constructs

For competition, it is normally most effective to combine multiple missions together into a "mission set" on the game board so that a single program can accomplish them before the robot returns to base. Once back in base, the two team members at the table reconfigure the robot by changing attachments and selecting a different program to execute the next mission set. Selecting the next program requires scrolling up or down to the correct program, which usually takes a few seconds and occasionally leads to an error (selecting the wrong program). Wouldn't it be nice if there was a way to avoid that error-prone selection process and save a few seconds at the same time?

Toward the end of our last season, my son, Graham, came up with a really cool idea we called the Combo program, so this section is dedicated to him. The concept uses a Switch block inside of a Loop block. The Loop block is in Unlimited mode, so once the program starts it runs continuously until stopped by the user. The Switch block is in mode Brick Buttons - Measure, which means it executes a different branch dependent on which button is pressed on the brick. Each mission program is converted to a My Block and inserted into a different branch of the Switch block. The default branch contains no blocks, so it does nothing but loop, waiting for the user to press a button on the brick. Because the brick has five buttons, this approach should work for up to five different mission sets.

I should note that we tried to implement this program near the end of the season, were not immediately successful, and decided to abandon it because we were so close to competition. It was not clear whether it did not work because we were doing something wrong or because the mission programs were so complex that the brick could not handle them being converted to My Blocks and executed inside a Switch block inside a Loop block. Theoretically, it should work fine, and the program I describe here worked well, but the My Blocks in this program are very small and simple compared to what the kids were running in competition. Nevertheless, if you want to try it, here's how.

First, convert your mission programs to My Blocks. Although they do not need inputs, it might be useful to create two or three "dummy" inputs for each one so the My Block is wide enough to display a useful name. Then, create a new program and name it Combo. Add a Loop block to start the program chain, and name it InfiniteLoop (or a similarly descriptive name). Place a Switch block inside the Loop block, and set its mode to "Brick Buttons – Measure." Click the "Add Case" button (see Figure 7-22) until there is one more case than the number of your mission sets. For example, five mission sets requires six cases in the Switch block. For each branch, click the indicator at the top and select a different button to activate it. Select the default button for the "no button press" branch. Finally, place a different mission program My Block into each branch of the Switch block. Leave the "no button press" branch empty. Your program should look something like Figure 7-23.

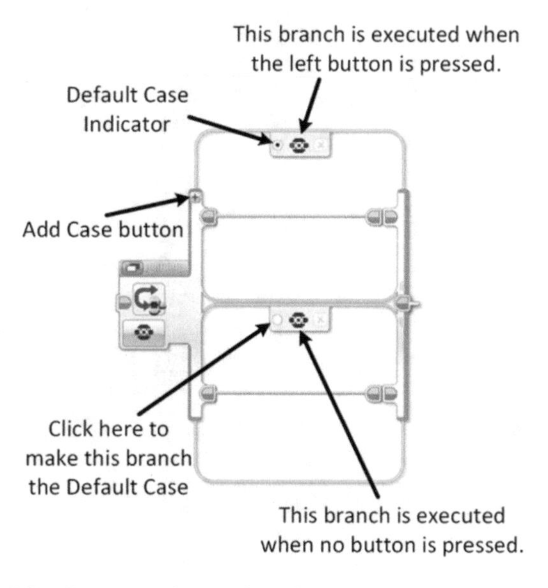

Figure 7-22. *Switch block controls and indicators*

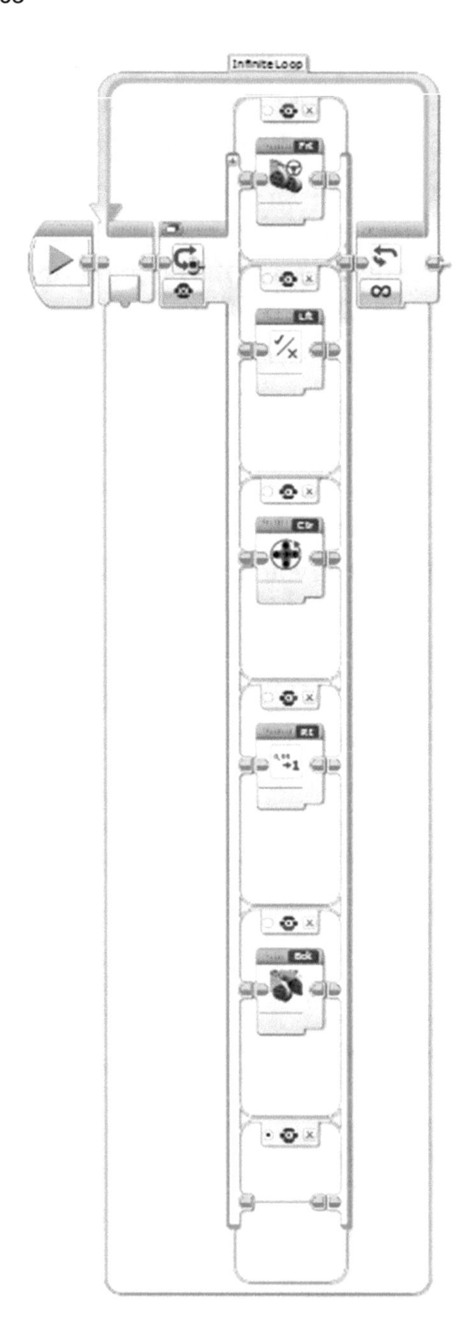

Figure 7-23. *Combo program*

Conclusion

If you have successfully implemented the My Blocks described in this chapter, you should have a powerful suite of My Blocks to help your robot navigate quickly and accurately on a game board. Next, the final chapter moves away from programming to discuss several topics related to building and managing a team, preparing for competition, and competing.

CHAPTER 8

Final Thoughts

Although good programming is important, FLL competition involves much more than good programming. This chapter begins with further considerations for competing effectively in the Robot Game, then expands to the judging rooms and general preparation during the season.

Robot Game

Prior to release of the Challenge, there are a few things that can be done to start preparing for the Robot Game. One of these is building a library of utility My Blocks, as described in the preceding chapters. A second task that can facilitate effective building of the robot and attachments is organizing the LEGO blocks and labeling the bins so everything can be found quickly and easily. A great deal of time can be wasted searching for parts if they are disorganized.

Another undertaking that must be completed is designing and building a robot. This is a good time to teach the kids about laying out requirements and using those in the design process. Typical requirements could include a low center of gravity; balancing narrow track width for navigating narrow passages against a wide track width for stability; durability; rigidity; low height to meet the 12" rule; and so forth. Be sure to consult the rubric when developing the requirements. If your team has more than one brick, you might want to have a robot design competition, and pick the winner using the Robot Design judging rubric. This activity will almost surely provide opportunities to talk about the Core Values. One caution is in order: If holding such a design competition after the season starts, limit it to a week or two. Otherwise, too much valuable time might be lost that should be used for programming and other activities.

G. Harding, *Programming LEGO® EV3 My Blocks*, https://doi.org/10.1007/978-1-4842-3438-9_8

Once the Challenge is released, it is essential to learn the rules. Ideally, everyone should learn them, although practically speaking it is probably fine if a couple of people know them well and everyone else has a general awareness of them. To encourage creativity, the general philosophy in FLL is usually "If it is not addressed in the rules, it is allowed." Although the organizers try very hard to address all of the details when they establish the rules, there are always questions. Thus, at least one person should be regularly checking for updates to make sure that a clarification or rule change does not cause your team a problem in the Robot Game.

It takes some time to analyze and understand the Robot Game each year. We found it effective to review each mission as a team, rating each one on difficulty (easy, medium, or hard) and point value while building a spreadsheet to document the analysis. We then selected easy missions with high point values to pursue first, and identified difficult and low-point-value missions to ignore. The other missions required more discussion and often involved more subjectivity, but some would be chosen, and others rejected.

Once the missions are chosen, group them into "mission sets," or groups of missions to be executed in a single autonomous run during the Robot Game. Finally, assign responsibility for each mission or mission set to one or two team members. If practical, it is a good idea to have a primary and alternate person assigned to each mission.

One thing I would probably do differently now is allow the students to program sounds on the brick for fun. This activity must be contained, however, or it could waste a great deal of time. Reasonable limits could include only one sound effect per mission, and only allowing it to be added after the mission programming is completed and the mission is successful. (We did not consider a mission successful unless the robot executed the mission correctly five consecutive times.)

Ensure all programs are commented, including My Blocks and mission programs. If done well, it will sometimes be very helpful during troubleshooting. It also makes the programming look more polished and professional, which helps in the judging room.

As the competition date approaches, there are a few things you can do to get prepared for the Robot Game. First, get a toolbox to store, transport, and protect the robot, charger, programming cable, attachments, and spare parts. We used a toolbox that was large enough to also hold the binders, folders, and smaller paraphernalia used in the judging rooms. Look for such toolboxes at home centers, department stores, and hardware stores in your area until you find one that suits your needs. Plastic toolboxes are generally lighter and less expensive than metal ones.

Second, a cart is very useful during competition, both for transporting all of the gear around the competition site, and as a staging platform at the game table. Just set the robot and all of its attachments on top of the cart, and roll it up to the game table. Although only two team members are allowed by the table at any given time, the cart is allowed to stay there throughout the round.

Third, if the Combo program (see Chapter 7) is not used, the Run Recent tab on the brick is very useful for quickly selecting and running a specific program. It is the leftmost tab on the brick display. Before going to the game table, run and quickly stop each mission program. This action puts each program on the Run Recent tab. If only those programs are on the Run Recent tab, it is fairly easy to find and run the one you need, saving valuable time at the table.

Finally, set aside one or two weeks at the end of the season to choreograph and rehearse the Robot Game. Although it takes discipline to stop work a week or two before competition day, the practice is crucial for reaching peak performance. Start by deciding the order of the mission sets and who will do them. Then walk through a full round, thinking carefully about where each person will stand and what they will do, when and how they will enter and exit, and so on. Once all of the issues have been worked out, try going through a couple of test rounds to see how long it takes and how much time can be shaved off by moving more quickly. Good choreography and rehearsal can often cut a minute or more off the time taken to complete a round.

The Judging Rooms

The Robot Game is arguably the most popular, and by far the most visible, part of an FLL competition. As discussed in the Preface, however, the Robot Game is really just a qualifier. Teams win or lose competitions in the judging rooms. Thus, it is paramount to understand how they work and prepare well for them. This section offers some tips to do that, starting with general guidelines that apply to all of the judging rooms, then offering a few specific strategies for individual judging rooms.

There are three judging rooms: Core Values, Project, and Robot Design. Each has a separate rubric used to judge the teams against nine skill areas, and each skill area has a description of what it takes to qualify as Beginning, Developing, Accomplished, and Exemplary. Study and learn the rubrics, and aim for the Exemplary level in each category. The rubric is your primary guide to prepare effectively for each judging room.

Keep in mind that the judges must evaluate many teams during the course of a competition. Although creativity is encouraged, do not assume that the judges will perceive some cool but subliminal implication of the team's skill. Make it explicit. Clearly address each skill area and explain how the team achieves the Exemplary level. (One caveat: Do not state that the team meets the Exemplary level; that sounds boastful. Just explain the things that the team believes make them exemplary and let the judges decide.)

Our teams tried to address the rubrics in two ways: first, verbally during the presentation time; and second, in writing with a one- or two-page summary that listed each skill area and how the team met it. Print several copies of the summary handout and give one to each of the three judges in the judging room. It is also wise to make extra copies. Larger tournaments might have three or more judging rooms operating simultaneously. It is common for top teams to receive callbacks to appear in front of all judges for tie breaking. Having extra copies of your handouts allows you to provide each judge with a copy during callback judging.

In the Robot Design judging room, it is required to have a printout of each program. We sometimes found the EV3 software a bit quirky when printing, so we usually took a screenshot of each program and printed that. If possible, a color printout is better because it is easier to read since the EV3 blocks are color-coded. As mentioned numerous times in previous chapters, every program should be commented because it makes it easier to understand and more professional in appearance.

Keep the first Core Value (we are a team) in mind for all of the judging rooms. One way to demonstrate this in the Robot Design judging room is to ensure each team member understands and is ready to explain at least one program or My Block to the judges. Although no one is expected to know all of the programs and My Blocks, at least one team member should be able to explain any particular program or My Block. If no one can explain a program, the obvious question would be whether the kids actually programmed it (Core Value 2, We do the work).

There is one thing we never tried that might be a good strategy in the Robot Design judging room: using photos of the robot to illustrate strengths in the mechanical design skill areas. Such photos could clearly indicate such things as how gussets and triangulation are used for strength and rigidity, how the brick and large motors are located down low for stability, how the charging port is left accessible for easy charging, or whatever your team wants to highlight regarding the physical construction of their robot.

Finally, I am a firm advocate of having a presentation ready for the Robot Design judging. It is not required, and occasionally there are judges who will not allow it, but giving a presentation makes the team look ready, and allows them to show off the best aspects of the robot. It also gives them something to practice in their preparations. The Robot Design judging normally includes about five minutes for the students to show off their robot and programs, then five minutes for Q&A, but it is good form for the team to ask the judges if it is okay for them to do a presentation and demo.

If your team is at a beginning or intermediate level, it is probably wise to demo a couple of the best, most reliable mission programs, and focus the rest of the presentation on programming and strategy, preferably with each team member having a speaking role.

If your team is advanced, has highly reliable programs, and really wants to take their presentation to a high level, they might want to try the technique described next. Although challenging, it looks very impressive if done well. Keep in mind that a Robot Game round lasts 2.5 minutes, and the "demo" time in the judging room lasts about 5 minutes, which means the robot action can be paused periodically in the judging room to allow more time for explanation.

First, make a copy of each mission program to be modified for the demo. Be sure to name each one something that clearly identifies it as a demo program, so it is not accidentally run during a real robot round. Then, insert "button-press" Wait blocks into the program at strategic points so that whoever is explaining the program has time to describe what is happening, after which he or she can press a brick button to let the robot continue. Finally, make sure each person on the team has a speaking role to describe some aspect of the robot's performance, and relates those comments back to something in the rubrics. It also might be effective to have one of your best speakers "emcee" the presentation.

The Core Values judging room is probably the least predictable because the team does not know what "team-building" challenge they will face. It usually begins with the team-building activity, some sort of challenge that generally requires problem solving, teamwork, and good communication to complete successfully. It is easy to find numerous such activities on the Web by searching "team-building activities." Pick a new one to practice each week, but do not tell your team what the activity is ahead of time. The element of surprise is an important part of the preparation.

Although solving the challenge is important, it is not nearly as important as the way the team interacts as they struggle to solve the challenge. The judges are looking for the kids to display the Core Values during this time, especially teamwork, Gracious Professionalism, and fun. It is possible to have a strong score in this judging room even if the team does not solve the challenge. Also, if you have one or more team members who are really shy, it might be a good strategy to pick some outgoing teammates, and assign them the responsibility to ensure the shy students participate actively by asking for their thoughts during the problem-solving phase.

Immediately following the challenge activity is a period of several minutes interacting with the judges. It is important that everyone on the team participate. As with the other judging rooms, I strongly recommend doing a brief presentation. Many teams use a trifold poster to illustrate the Core Values. A good practice is to get at least one picture of each student practicing the Core Values in a variety of circumstances. Some students might use a picture showing a robotics activity, whereas others illustrate activities with friends, and still others depict an activity with family. The point is to show that the Core Values are being implemented and practiced both within and outside of FLL activities.

I recommend asking the judges if it is okay to do a brief presentation, then having each student explain the story behind his or her photo, and name the Core Value it illustrates. Practice the presentation in the weeks leading up to competition.

We also assembled a small portfolio for the Core Values judging room. The cover page had our team name and team photo, followed by the official FLL Team Information Sheet. We also included the following documents:

- Core Values Rubric Summary: This listed each skill area and what the team did to meet it.
- Core Values Summary: This was an additional page of text the team wanted in the portfolio. It was a three-paragraph narrative of how the team managed their activities during the season.
- Team Rules: Our team had a set of rules that extended the Core Values in practical ways, such as punctuality and cleaning up after themselves.
- Team Goals: The team set a number of goals at the beginning of the season, so they were listed here.

- Core Value Exercise Summaries: This section was several pages long. Each page had a description of the activity, how the team did at it, the debriefing discussion afterward, lessons they learned and the Core Values that applied, and a photo of the team engaged in the activity.

A good portfolio with pictures can really bring to life what the kids did to prepare for Core Values judging. Leave at least one with the judges, and preferably leave one with each judge.

As with the other two judging rooms, use the rubric as your guide to Project judging. As the kids do their research, tell them to save links and print out good information. Help them organize the information into a binder. Ours was about an inch thick by the end of the season. This makes a great visual to indicate how much time and effort the team expended doing their research.

A summary portfolio is also an excellent thing to have for the Project. It can include a number of things, including these:

- Cover page.
- FLL Team Information Sheet.
- Summary of how the team met the rubrics.
- Problem statement.
- Proposed solution.
- Overview of research done.
- Description of interactions with experts and professionals.
- List of references.
- Skit script.

The Project judging room typically includes a one-minute setup time, during which the team carries in their props and materials and sets up for the skit, followed by five minutes to perform the skit, and then a five-minute Q&A period with the judges.

Even if the skit is well done and the kids speak clearly, there is often room for someone watching the skit to misunderstand or misinterpret portions. Something we tried in our last season that seemed to work well was using posters to emphasize key points, such as scene locations and rubrics. For instance, in the middle of the skit, a student who was not in that particular scene would hoist a poster with a brief phrase, such as "Alert: Solution," and say the phrase aloud to call attention to it, so it was clear to the judges what was happening.

Note that in all of the judging rooms the coaches are not allowed to interact with or help the team. In some competitions the coaches might not even be allowed into the judging rooms, but if they are allowed in the rooms, they are expected to be quiet and unobtrusive. This makes practice critical. In the last couple of weeks or so before competition, script and practice the Project skit, the Robot Design presentation, and the Core Values presentation. Hold mock judging sessions where you try to simulate real competition conditions as closely as possible, including asking the kids tough questions and interrupting their presentations. These things happen in competition sometimes, and it is better for the kids to face them for the first time in practice rather than at a competition. Remember the old adage: You play like you practice!

General Notes

A building is only strong and stable if it rests on a strong foundation. Generally speaking, teams only compete well if they have laid a strong foundation of planning and preparation in the weeks and months leading up to the competition(s). This final section of the book presents some ideas for laying that strong foundation with your teams.

First, set team goals early and make sure everyone clearly understands the goals and the commitment required to achieve them. It is quite possible that some kids will expect FLL to be little more than organized playtime, whereas others might want to strive for a state championship. Likewise, some parents might view it as free babysitting for their children, whereas other parents are willing and able to provide substantial volunteer time to enable the team's success. It is virtually impossible to harmonize such disparate views and expectations, so it is very important to clarify expectations early. In some cases, it might mean that some kids do not participate. In other cases, it could mean two separate teams, which is okay. It is certainly acceptable not to strive for a high degree of competitiveness, but to focus on having fun and learning a little along the way. The problem occurs when members of the same team have widely differing goals and expectations.

The remainder of this chapter contains tips and advice for preparing to be competitive. It is up to you and your team whether and how to implement them, as appropriate for your team.

Planning practices ahead of time is generally wise. Adjusting the schedule on the fly is often necessary, but having a schedule helps ensure time allocated to each area: Robot Game and Judging, Project, and Core Values. This facilitates good preparation and is something the judges often ask about (spending equitable time in each area).

I strongly recommend spending time every day learning the eight Core Values. We started with the first two or three, then added one or two each time we met. I quizzed the team, asking each player in turn to recite one of the values. Once we had all eight memorized in order, I quizzed them out of order (e.g., "Graham, what is Core Value number four?", "Jackson, what is Core Value number seven?", etc.). The repetition embeds the values into everyone's memory, which facilitates their application. It is also impressive when students can recite the Core Values in front of the judges in the judging rooms.

Leading a successful FLL team is a lot of work. Do not try to do it alone. Recruit assistant coaches in each area, especially where you might lack some skills. If you are the head coach, you will likely lead one of the three areas (Robot Game, Project, or Core Values), so try to find parents or other adults to take the lead in the remaining areas. Also, recruit someone to coordinate snacks. Having one or two snack breaks each practice is very important. All of the parents can take turns bringing snacks, but someone must coordinate the effort and maintain the schedule. This is a relatively easy task, and might be perfect for a parent who, for whatever reason, is not up to providing regular assistance to the team. In my experience, one break was usually sufficient for older kids, whereas younger kids sometimes need multiple breaks, especially for longer practices.

Normally it is not a good idea to focus on the same activity for an entire practice. Rotate the kids from one task to another at least once or twice each practice. Changing activities reduces boredom and tends to keep the kids more engaged.

On a similar note, I highly recommend assigning each team member a role in every area (Core Values, Project, and Robot Game). If we are not careful, it is easy to lose the big picture when aiming for a high level of competitiveness, but the bigger picture is that we are helping prepare the kids for *life,* not just to compete effectively. Think about all of the things our kids are learning:

- Research.
- Writing.

- Presentation skills.
- Programming, and the logic that comes with it.
- Building, and the physics principles to go with operating a robot.
- Teamwork.
- Troubleshooting and critical thinking.
- Maintaining composure under pressure.
- Competing hard while encouraging their competitors.

A given student might be a really good writer, but a weak programmer, so it makes sense to use his strength in documenting Project work. Nevertheless, he should have some role in the Robot Game, even if it is minor. Likewise, strong programmers and builders could work on more complex missions in the Robot Game, but should still have roles in other areas, even if they are relatively minor. We are trying to prepare these kids for life, not just to win a competition.

When you talk to others about FLL, be sure to sell and explain the program. Most people who have never been involved have no idea what it involves and the incredible benefits to be obtained. Tell them!

Finally, thank you for what you are doing to help your students. FLL cannot work without the vast array of people who volunteer to be coaches, mentors, judges, tournament organizers, and so on. Never forget that you are doing a great thing: Thank you!

Glossary

Dead reckoning: A method of navigation in which one's position is determined by adding the distance and direction traveled to the starting position. Position error in this type of navigation tends to get larger as the distance traveled increases.

Error: The difference between the desired and actual values of a parameter being controlled in a control system.

EV3: The third generation of the Mindstorms robot hardware and software.

Feedback: Information about the operation of a system, used for either control, or monitoring and troubleshooting.

- **Control:** In the context of a robot controlling its own operation, such as following a line, one or more sensors are used to provide information about system operation, then that information is compared to the desired operation to create an error signal, which is used to adjust system operation.
- **Monitoring and troubleshooting:** Sometimes it is useful to have the robot provide information back to the user running the robot (e.g., "line detected") to ensure the robot is really doing what it appears to be doing, and to assist with troubleshooting.

Firmware: Quasi-permanent software stored in read-only memory and designed to facilitate interaction with a computing system's hardware.

Friction: A physical quality that resists sliding or slipping motion between two surfaces.

G. Harding, *Programming LEGO® EV3 My Blocks*, https://doi.org/10.1007/978-1-4842-3438-9

Geometry: A branch of mathematics that deals with the properties of shapes like lines; angles; triangles, squares, rectangles, and other polygons; circles and ellipses; and so on.

Handoff: Leaving the motors running at the transition from one My Block to the next, saving time by not stopping and restarting the robot.

Loop: A program construct that executes one or more commands repetitively. A loop can execute a specified number of times, execute until a condition is met, or run until the user terminates the program (an infinite loop).

Modularity: The principle of dividing programs into smaller, more manageable, and reusable blocks.

My Block: A subprogram, or subroutine, that is designed to perform a repetitive task or set of tasks. Such a subroutine is programmed, debugged, and tuned once, then used multiple times in other programs.

Parallel execution: Running two independent sections of software simultaneously. This can be done with the EV3 by plugging the output of a block into the inputs of two or more subsequent blocks.

pi (π): A math constant used to, among many other things, convert between the diameter and circumference of a circle. Its value is approximately 3.14. The number extends to an infinite number of decimal places, 3.1415926..., but for our purposes 3.14 will suffice.

Proportional controller: A type of controller that adjusts the amount of correction in direct relation to the amount of error; a small error, or deviation from the desired value, results in a small correction to move closer to the desired value, also called the set point; a large error generates a large correction to move more quickly toward the set point.

Sequence plug entry: The point in a programming block that connects to the previous block.

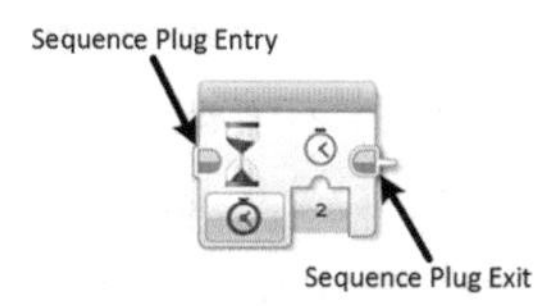

Sequence plug exit: The point in a programming block that connects to the next block.

Set point: The desired value, such as distance or reflected light intensity, of a system being controlled.

SONAR: Acronym for SOund NAvigation and Ranging. This technique measures the time it takes for a transmitted sound wave to travel to an object, reflect off of it, and return; it then uses that time to compute the distance to the object.

Torque: A physical quantity that tends to create rotation. For instance, a motor produces torque to rotate its shaft, which could be attached to a gear or wheel and create movement. A lever attached to a shaft can be used to create torque on the shaft, like a wrench can be used to create torque on a bolt.

Track width: Center-to-center distance between the two drive wheels of the robot.

Traction: In the context of a wheel or tire on a robot, traction could be described as the friction between the tire and the surface on which it is rolling. This friction allows the tire to exert force on the surface to move the robot.

Unit conversion: Translation of a value from one type of measurement to another (e.g., converting 1 inch to 2.54 centimeters).

Variable: An element used to store a value that can be changed. In EV3 programming, the Variable block is under the Data Operations tab. It is set to Read mode when its value is to be passed to another element and Write mode when the value it contains is to be replaced with a new value. A Variable block can be any one of the following data types: Text, Numeric, Logic, Numeric Array, or Logic Array.

Waypoint: A location with precisely known coordinates that can be used for correcting navigation position.

Index

G. Harding, *Programming LEGO® EV3 My Blocks*, https://doi.org/10.1007/978-1-4842-3438-9

E

F

G

H

I

J, K

L

M

N

O

P, Q

R

S

T

U

V

W, X, Y, Z

Printed in the United States
By Bookmasters